Wireless Networks

Series Editor

Xuemin (Sherman) Shen, University of Waterloo, Waterloo, Canada

The purpose of Springer's Wireless Networks book series is to establish the state of the art and set the course for future research and development in wireless communication networks. The scope of this series includes not only all aspects of wireless networks (including cellular networks, WiFi, sensor networks, and vehicular networks), but related areas such as cloud computing and big data. The series serves as a central source of references for wireless networks research and development. It aims to publish thorough and cohesive overviews on specific topics in wireless networks, as well as works that are larger in scope than survey articles and that contain more detailed background information. The series also provides coverage of advanced and timely topics worthy of monographs, contributed volumes, textbooks and handbooks.

Li-Chun Wang • Swades De •
Ashutosh Balakrishnan

Green Energy Driven Integrated Smart Grid and Wireless Networks

Springer

Li-Chun Wang
Department of Electrical and Computer Science Engineering
National Yang Ming Chiao Tung University
Hsinchu, Taiwan

Swades De
Department of Electrical Engineering and Bharti School of Telecommunications
Indian Institute of Technology Delhi
New Delhi, Delhi, India

Ashutosh Balakrishnan
Department of Computer Science and Networks
IMT-Télécom Paris, Institut Polytechnique de Paris
Palaiseau, France

ISSN 2366-1186 ISSN 2366-1445 (electronic)
Wireless Networks
ISBN 978-3-032-20508-7 ISBN 978-3-032-20509-4 (eBook)
https://doi.org/10.1007/978-3-032-20509-4

This Springer imprint is published by the registered company Springer Nature Switzerland AG
The registered company address is: Gewerbestrasse 11, 6330 Cham, Switzerland

"When wireless is perfectly applied, the whole earth will be converted into a huge brain, which in fact it is, all things being particles of a real and rhythmic whole. We shall be able to communicate with one another instantly, irrespective of distance. Not only this, but through television and telephony we shall see and hear one another as perfectly as though we were face to face, despite intervening distances of thousands of miles; and the instruments through which we shall be able to do his will be amazingly simple compared with our present telephone. A man will be able to carry one in his vest pocket."

Nikola Tesla (1926)

Foreword

Sustainability has rapidly moved from being a desirable attribute of communication systems to a central pillar of next-generation network design. As we look ahead to sixth-generation (6G) communications—an era characterized by ubiquitous connectivity, unprecedented data demands, and stringent security requirements—energy efficiency and environmental stewardship are no longer optional. They are essential. This monograph arrives at precisely the right moment, offering a comprehensive and forward-looking treatment of green, resilient, and intelligent communication infrastructures.

At its core, this work reimagines the relationship between wireless networks and the power grid. Traditional power systems, though robust, were never designed to serve the dynamic, high-density, and energy-intensive demands of modern wireless communication. Likewise, existing standalone green communication systems often lack the scalability required to support large-scale deployments. By proposing an integrated architecture that couples renewable energy sources with the conventional grid—and by positioning each grid user as a potential energy contributor—the monograph advances a transformative vision: a distributed, ambient-powered network that operates not only as an energy consumer but also as a carbon-free prosumer.

The book's structure reflects both intellectual rigor and pedagogical clarity. Early chapters (Chaps. 1 and 3) provide beginners with the motivation and foundational insights needed to understand sustainability in 6G systems. They map the evolution of research themes, outline the challenges associated with system design, and introduce key cost metrics that govern energy-aware networks.

Chapter 2 offers a benchmark study of solar-powered, off-grid base stations—an increasingly relevant component of modern telecommunication ecosystems. Through a careful Markovian analysis, the chapter provides first-principles tools to quantify outages in ambient-powered nodes, establishing a solid foundation for subsequent chapters.

Chapters 4 and 5 take the discussion deeper, providing a rigorous analytical framework for grid-networked and ambient-powered telecommunication systems. These chapters illuminate how cooperative energy balancing and energy trading can dramatically enhance network sustainability. From both user-centric and operator-centric viewpoints, the analyses show how green energy utilization, service reliability, and revenue considerations intersect—resulting in more scalable and economically viable network architectures.

Chapter 6 explores hybrid aerial-terrestrial networks, a topic of growing interest as industry and academia investigate the possibilities of aerial base stations. By studying the intrinsic coupling between data and energy networks, the chapter demonstrates how aerial platforms can augment terrestrial systems in resource-constrained environments. Chapter 7 broadens this perspective further by extending energy balancing concepts to residential settings, demonstrating their relevance to IoT, sensor networks, and even green data centers.

The monograph culminates in Chap. 8 with a compelling exploration of "green AI" in wireless communication systems. As AI becomes integral to future networks, understanding the energy–performance trade-off is vital. The chapter outlines emerging directions for AI-assisted network optimization while emphasizing the urgent need for computationally lightweight, energy-efficient models.

The depth and breadth of the work reflect years of meticulous research, rigorous analysis, and a strong commitment to advancing sustainable communication technologies. It will undoubtedly serve as a valuable resource for undergraduate and graduate students, academic researchers, and industry practitioners alike. As the world moves toward greener, more intelligent communication infrastructures, the ideas presented in this monograph will help shape the next generation of sustainable network design. It is my pleasure to commend this important contribution to the community.

Winnipeg, MB, Canada
December 2025

Ekram Hossain

Preface

This monograph outlines the design of green energy-driven, integrated smart grid and wireless networks. The need for the monograph stems from increased power consumption in contemporary communication systems, the inflexibility of the traditional power grid, and the non-scalability of standalone green communication systems.

As an advance, the monograph explores the integration of renewable energy with the traditional power grid. The core idea is to enable each grid user to be a potential energy source for the grid, thereby revising the conventional idea of power grids. A grid networked system of such distributed ambient powered nodes can potentially act as a carbon-free energy producer system to the power grid in addition to an energy prosumer system.

The monograph is expected to be very beneficial to the researchers and engineers, with scope varying from undergraduate to graduate students. For instance, the beginners may motivate themselves toward green communication from Chaps. 1 and 3. Chapter 1 outlines the upcoming research theme of 6G communications and highlights the importance of sustainability. From a beginner's perspective, Chap. 3 presents insights on the design constraints and challenges in system analysis. It details the inherent cost metrics and network operations associated with ambient powered tele-networks.

Chapter 2 acts as a benchmark, wherein a detailed design and performance evaluation in purely off-grid solar powered base stations is presented. The chapter includes insights on the outage prone-ness of ambient powered nodes, and presents a Markovian framework to estimate the outages through first principles.

Chapters 4 and 5 present an in-depth analysis of grid networked and ambient powered telecommunication networks, presenting optimal amounts of energy which the nodes can share among each other or trade with the grid. Chapter 4 first presents a cooperative energy balancing based framework among the networked nodes. Through energy balancing, the improvement in network green energy utilization is shown, thereby realizing carbon-free green networks. Chapter 5 analyzes the ambient powered and networked system from an operator perspective. By incorporating revenue awareness, it illustrates the beneficial effects of energy balancing and its complementarity to load balancing, aiding the scalability of green solutions from an industry perspective. From a system perspective, a joint load-energy balancing is studied, showing insights on the operator revenue being influenced by user service and network green energy utilization.

Chapter 6 showcases an aerial-terrestrial green network design, wherein the inherent coupling between the data and energy networks of the system are studied. As a case study, system performance is evaluated for an aerial base station (BS) aided terrestrial network. Chapter 7 demonstrates a wider perspective of the energy balancing in networks, through a residential setting. It is worthwhile to note that the concept of energy balancing can be easily extended to energy aware network design, including ambient internet of things (IoT), wireless sensor networks, or green data center networks.

Chapter 8 presents a glimpse of green artificial intelligence (AI) aided future networks. The key research directions wireless communication networks are showcased, wherein AI can be used to improve upon the current state of art. The chapter also gives a detailed discussion on upcoming green AI models, the trade-off between energy efficiency and AI performance, and the need to develop computationally light and efficient AI models for future networks.

The material presented in this monograph originates from the work done by Ashutosh Balakrishnan for his Ph.D. jointly at IIT Delhi and NYCU Taiwan. The research disseminated is expected to be very useful not only for academic researchers but also to industry in scaling green service provisioning solutions to large networks.

Hsinchu, Taiwan | Li-Chun Wang
New Delhi, India | Swades De
Palaiseau, France | Ashutosh Balakrishnan
December 2025

Declarations

Competing Interests The authors have no competing interests to declare that are relevant to the content of this manuscript.

Ethics Approval The authors declare that no part of this monograph contains study on humans. Furthermore, it is also declared that no human participant or animals were involved in any of the studies contained in this monograph.

Contents

Acronyms

4G	Fourth generation
6G	Sixth generation
5G	Fifth generation
aBS	Aerial BS
AC	Alternating current
AI	Artificial Intelligence
BS	Base station
CAPEX	CAPital EXpenditure
CDF	Cumulative distribution function
DC	Direct current
DTMC	Discrete time Markov chain
EC	Energy cooperation
eMBB	Enhanced mobile broadband
FCC	Federal communications commission
GEB	Green energy balancing
GEO	Geostationary earth orbit satellite
HAPS	High altitude platform stations
hMBS	HAPS mounted macro BS
ICT	Information and communication Technology
IoT	Internet of Things
ISAC	Integrated Sensing and Communications
JTEB	Joint traffic-energy balancing
LEO	Low earth orbit satellites
MA-RES	MAcro grid connected REsidential System
MDP	Markov decision process
MEO	Medium earth orbit
MI-RES	MIcro grid connected REsidential System
mMTC	Massive machine type communications
MSE	Mean squared error
NREL	National Renewable Energy Laboratory
NTN	Non terrestrial networks

OEMC	Operations and energy management center
OPEX	OPerational EXpenditure
PDF	Probability density function
PV	Photo voltaic
QoE	Quality of experience
QoS	Quality of service
RAN	Radio access network
RIS	Reflective intelligent surfaces
RMSE	Root mean squared error
SAGIN	Satellite aerial ground integrated networks
SINR	Signal interference to noise ratio
SNR	Signal to noise ratio
TLB	Traffic load balancing
t(M)BS	Terrestrial macro BS
UAV	Unmanned aerial vehicles
UE	User equipment
URLLC	Ultra reliable and low latency communications
USD	United states dollar
WEC	Without EC

List of Symbols

$\mathbb{E}(\cdot)$	Expectation operator
$var(\cdot)$	Variance operator
$\mathbb{P}(\cdot)$	Probability operator
$\mathbb{Q}(\cdot)$	Tail distribution function of the standard normal distribution
$f_X(\cdot)$	PDF of random variable X
$\mathbb{L}(\cdot)$	Lagrangian operator
A	Considered area of study
b	BS index, denotes BS b
t	Time index
$\mathcal{H}_b(t)$	Temporal energy harvest of BS b
β^{max}	Upper limit of battery capacity
β^{cr}	Critical limit of battery storage
$\beta_b(t)$	Temporal battery level of BS b
$\mathcal{E}_b(t)$	Temporal power (i.e., energy) consumption of BS b
$L_b(t)$	Temporal intensity of traffic load at BS b
$\mu_L(t)$	Average load intensity at time t
σ_X^2	Variance of random variable X
N_{trx}	Number of radio transceiver per BSs
$\mathcal{P}_0$	Static power consumption of a BS, irrespective of BS temporal load
$\mathcal{P}_{max}$	Maximum downlink power budget mandated by FCC
Δ	Power amplifier efficiency
P_{ij}	Transition probability from state i to state j
$\underline{T}$	Transition probability matrix
$\underline{\pi}$	Steady state probability vector
π_i	Steady state probability of state i
P_n	Outage probability for n consecutive time slots
$\mathbb{D}_b(t)$	Difference energy (difference of temporal energy harvest and BS load)
$\mathcal{R}$	Rating of PV panel
$\mathcal{L}_{PV}$	Lifetime of PV panel
$\mathcal{Y}$	Time frame of analysis (in years)
$\mathcal{B}$	Number of BSs in the network

δ	Depth of discharge
$\mathcal{L}_B$	Lifetime of storage battery
N_{PV}^O	Optimal PV panels in off-grid solar BS
$\mathcal{N}_B^O$	Optimal number of batteries in off-grid solar BS
C_{PV}	Cost of unit PV panel
C_B	Cost of unit storage battery
$\mathcal{B}_c$	Total storage capacity with a BS
λ	User density
$\Gamma_b(t)$	Skewness/burstiness of the load intensity experienced by a BS b
ζ	Parameter to quantify the degree of traffic burstiness experienced at the BS, termed as skewness factor
ES	Number of energy-sufficient BSs, indexed by b'
ED	Number of energy-deficient BSs, indexed by b
$\mathcal{D}(b, t)$	Amount of energy deficit at BS b at time t
$\mathcal{E}^S(b', t)$	Amount of sharable energy with BS b' at time t
$\mathcal{G}_b(t)$	Amount of grid energy procurement by BS b at time t
$\mathcal{S}_b(t)$	Amount of energy shared to BS b by the energy-sufficient BSs
C^{ser}	Per day revenue earned by a mobile operator by serving a user
C^{sh}	Cost of sharing unit energy among the networked BSs through the power grid
C^{sell}	Cost of selling unit energy to the power grid by the BSs
C^{buy}	Cost of purchasing unit energy from the power grid by the BSs
BW_b	Bandwidth resource available with BS b
r_{ub}	Data rate achievable by user u when associated with BS b
SNR_{ub}	Signal to noise ratio experienced at user u when associated with BS b
$\mathcal{P}_{ub}(t)$	Power allocated to user u by BS b
g_{ub}	Channel gain of user u when associated with BS b
d_{ub}	Distance of user u from BS b
θ_{th}	SNR threshold, denoting user QoS

List of Figures

List of Tables

Chapter 1
The Need for Sustainable Communications

1.1 Introduction

The Information and Communication Technology (ICT) sector has witnessed a continual technology evolution from purely analog communication to the upcoming digital communication based sixth generation (6G) communications [3]. 6G is aimed to build on the existing fifth generation (5G) communication technology as illustrated through Fig. 1.1 [4, 5]. The current 5G communication technology has been developed specifically for realizing the following use-cases:

- ultra reliability and low latency communications (URLLC)
- massive machine type communications (mMTC)
- enhanced mobile broadband (eMBB) services

These use-cases are expected to be crucial for applications like virtual reality based services, industry automation, autonomous vehicles, tactile internet, robotic arm applications in medical services, etc. [6–13]. The massive push in mobile connectivity through current 5G communication technology has led to exponential increase of network users, alongside increasing per user quality of service (QoS) demands [14, 15]. Recent estimates indicate a rise in the number of mobile subscribers from 5.8 billion (in 2024) to around 6.5 billion globally, by 2030 as shown in Fig. 1.2. The adoption of 5G communications by the subscribers is also encouraging and is expected to witness a stupendous increase from current 25% to around 57% by 2030 [16].

To cater to the rising user QoS and to further improve the network performance, the 6G communication is envisaged to advance these 5G verticals further through,

- *immersive communications* or semantic communications, i.e., communicating only data that is necessary thereby saving bandwidth and making the communication more spectral efficient.

L.-C. Wang et al., *Green Energy Driven Integrated Smart Grid and Wireless Networks*, Wireless Networks, https://doi.org/10.1007/978-3-032-20509-4_1

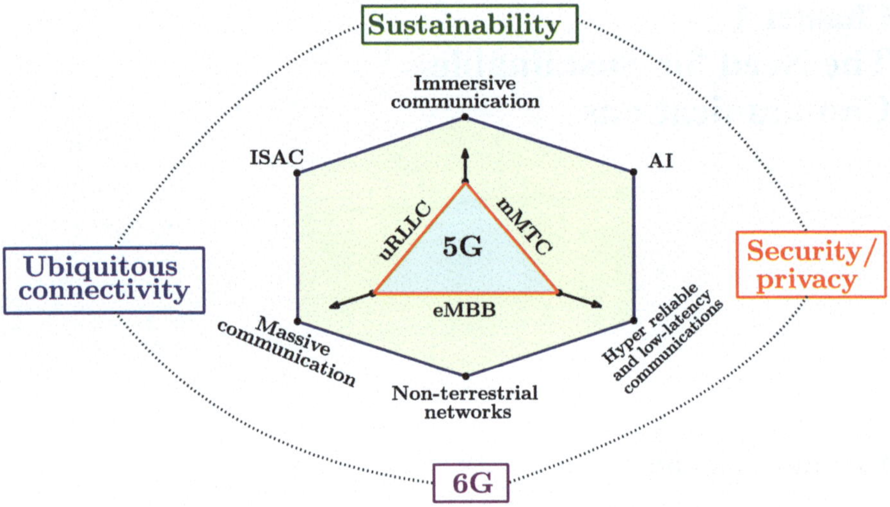

Fig. 1.1 Research directions of upcoming 6G communications

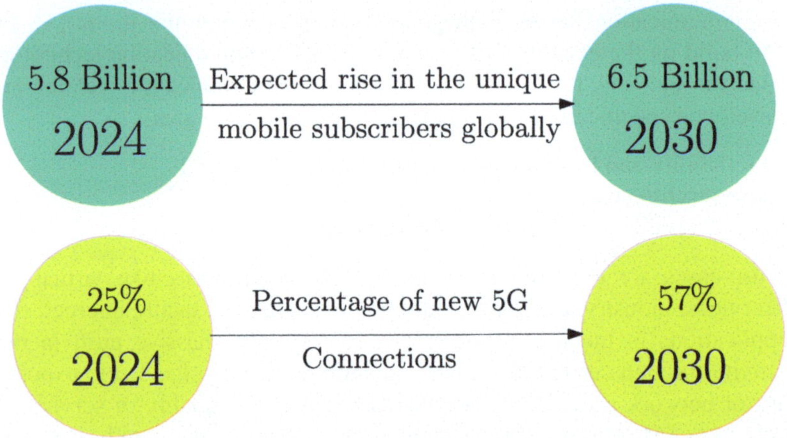

Fig. 1.2 Estimated increase in number of mobile subscribers and the percentage adoption of 5G by 2030, globally

- *integrated sensing and communications (ISAC)* based network architectures wherein the mobile BS can perform sensing as well as communication operations simultaneously [17].
- *massive cell free communications*, wherein the idea is to go beyond "cells" in a communication network.
- provide ubiquitous connectivity by integrating existing communication infrastructure with *non-terrestrial networks (NTNs)* like unmanned aerial vehicles (UAVs), high altitude platform stations (HAPS), or satellite networks.

- creating robust automation to improve performance efficiency of networks through *artificial intelligence (AI)*.

These upcoming verticals of the 6G technology are encompassed with three key challenges,

- Ubiquitous connectivity
- Secure network design
- and, Sustainable network design.

Ubiquitous connectivity is challenging as it aims to provide telecommunication connectivity to farthest and rural areas, connecting people. It also aims to connecting different technologies, aiming inter-operability through open radio access network (RAN) design. The continually increasing user QoS and quality of experience (QoE) requirements to the mobile user (gauged generally in terms of data rate guaranteed by the mobile service provider), is a key enabler for the 6G verticals discussed above. Key research for ubiquitous intelligence and connectivity is expected to move towards integrating the existing communication architecture with AI for better performance [18–20], exploring new meta-surface based infrastructures to aid communication performance through reconfigurable intelligent surfaces (RIS) [21–24], and explore the possibilities of aiding the existing infrastructure with aerial and space based NTNs [25–27].

Achieving secure communications is critical, considering the advancements in computational technologies including quantum communications. With most of the human-area-network around us becoming "digital", secure communications becomes more critical in areas like banking sectors, digital communication platforms including emails and mobile applications, digital payments, etc., directly affecting day-to-day life. Among nations, cyber-warfare is being increasingly used to paralyze the digital infrastructures especially in war times, thereby stressing the need for secure networks. While we acknowledge the need for secure network designing, it is beyond the scope of this monograph and is left for future research.

Sustainability is the third challenge in realizing 6G networks. We strongly emphasize that the term "sustainability" should not be linked with purely energy sustainability. While sustenance is seen from the energy perspective, economic sustainability directly affects scalability thereby limiting ubiquitous connectivity, and hence the theme of this monograph.

The upcoming sections motivate the objectives of this monograph.

1.2 Motivation

The primary motivation behind the monograph stems from increased power consumption of 5G and beyond systems, inflexibility of the traditional power grid and nonscalability of standalone green communication systems, thus motivating to potentially design *networked ambient powered intelligent communication systems* by leveraging the grid infrastructure. In the upcoming subsections, the motivation towards the monograph is presented in detail.

1.2.1 Increased Power Consumption in 5G and Beyond

A wireless RAN primarily consists of user equipments (UE), base stations (BSs), and the core network. From a technical perspective, the data rate achievable by a user is governed by controllable parameters and stochastic factors. The controllable parameters include the physical resources, namely, bandwidth spectrum and power allocated by the BS to the concerned user, as has been established by the Shannon-Hartley theorem [28]. The stochastic factors include wireless fading channel and the additive noise which gets added at the receiver hardware. Optimizing the radio resource allocation by factoring in the stochastic factors does lead to energy savings, but the magnitude energy savings is not expected to be significant, owing to the continual addition of UEs to the network, and the increasing UE QoS demands.

At the network level, BS densification, i.e., increasing the deployments of BSs in the network, is the most effective way to cater to the increasing network users and QoS demands by the mobile service provider [29–32]. Recent industry reports have estimated that with adoption of 5G communication technology, the system capacity is expected to rise approximately 100 times. Furthermore, millimeter (mm) wave transmission technology is expected to increase the cell density exponentially [33] by approximately 20% till 2030.

With frequency being a scarce resource, increasing network users and user QoS implies a direct increase in the power allocation to the user thereby increasing the BS power consumption. The BS is the most energy intensive component in a communication system, consuming up to 58% of the energy consumed by a communication system as illustrated in Fig. 1.3 [34]. A significant part of the BS energy consumption is at the radio access network through: RAN antennas, radio

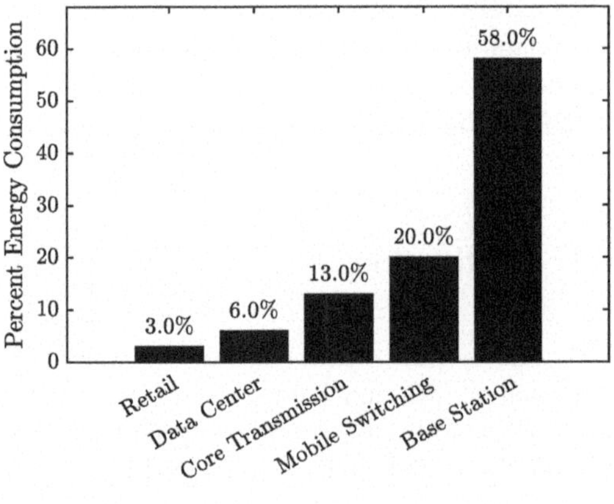

Fig. 1.3 Energy consumption by various components in ICT sector

units, base station elements, core/edge cooling, computing processes, amplifiers, and backhaul units. A standalone diesel powered BS is estimated to consume about 1500 liters of diesel per month, generating around 4000 kg of carbon di-oxide (CO_2). Furthermore, as compared to the fourth generation (4G) systems, a 5G BS is expected to consume at least twice the amount of energy as compared to a 4G BS system [33].

With the ICT sector roughly consuming around 10% of global electricity [35], network densification towards ubiquitous connectivity is expected to result in an estimated increase in network energy consumption by 170% in the upcoming years [29]. Hence, "energy sustainability" becomes a critical challenge with rise in ubiquitous connectivity. Therefore, it is imperative to devise energy efficient networking frameworks, to offset/ mitigate the increased energy savings. Integrating renewable energy with the existing grid might be a very attractive solution [36], as has been discussed in this monograph.

1.2.2 Inflexibility of Traditional Power Grid

Traditionally, the BSs have been powered through the power grid [37, 38], which operates through carbon emitting coal based thermal power plants. There have been multiple global instances recently related to power grid failure leading to complete electricity blackout [39]. The over-reliance of traditional power grids on carbon generating thermal power plants makes the power grid unreliable. Additionally, there is a lack of communication between the grid users (demand side) and the power grid utility company (supply side), resulting in frequent supply-demand mismatch. It may be noted that grid users may refer to cellular BSs or residential systems which are connected to the power grid. Since this monograph specifically focuses on cellular systems, hence grid users will implicitly refer to cellular BSs unless stated explicitly. Due to such a lack of flexibility and unreliability in the traditional power grid, there is requirement to upgrade the power grid to make it a more reliable, energy efficient, and responsive system. Intelligently handling the supply-demand mismatch is not only beneficial in balancing the load of cellular system, thus improving user QoS, but also has the potential to provide cost benefits to the mobile service provider (who operates the grid users, i.e., BSs). Additionally, perceiving the significant threat of global warming and climate change, it is imperative that the ICT sector also contribute novel strategies to mitigate the carbon footprint.

1.2.3 Non-scalability of Standalone Green Systems

A very effective solution to cut down the carbon footprint, is to power the grid users, i.e., BSs with purely with renewable energy sources like, solar [40], wind

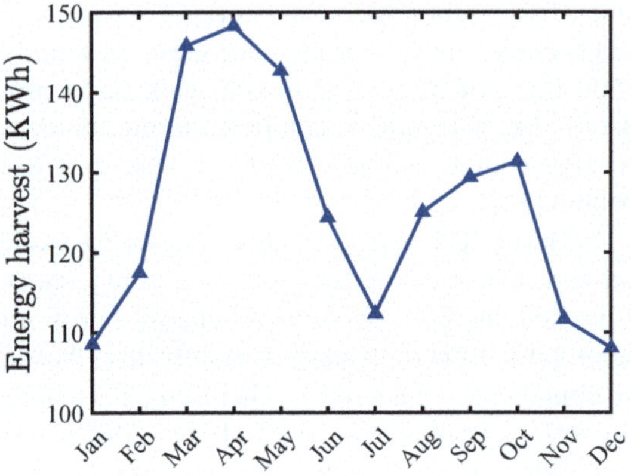

Fig. 1.4 Monthly variation in energy harvest by a 1 KW photovoltaic panel at New Delhi city, India

[41], or geothermal energy [42, 43], thus making the grid users standalone and off-grid. In solar abundant countries like India, which is sitting on a cusp of telecommunication revolution, solar energy powered BSs can be a very effective strategy to cut down carbon footprint and thus move toward green communication [44]. Powering BSs purely with renewable energy sources like solar or wind energy comes with a different set of challenges. These renewable energy sources are prone to climatic changes, i.e., the energy harvesting efficiency is dependent on the weather conditions. For instance, solar energy harvest will be very effective in summer and spring seasons but the harvesting efficiency will be very poor in rainy and fog based seasons, as illustrated in Fig. 1.4. Renewable energy like wind energy cannot be generated in plain land areas and is more prominent in coastal regions. Solar energy in particular cannot be harvested in the night time, hence the system requires to be provisioned to additional battery supplies to supply the system when the harvest is not good, thereby preventing energy outage. Hence, powering the BSs purely with ambient power supplies poses extra capital expenditure (CAPEX) to the mobile operator [45], and is not a scalable strategy.

Cost or revenue to the mobile network operator has emerged as a very crucial parameter alongside energy efficiency, so as to realize scalable networks. For instance, BS densification leads to significant deployment costs to the mobile operator, in addition to compounding the network energy consumption as discussed in Sect. 1.2.1. There is a lacuna in the current literature to study the various costs associated with the telecom operator necessitating a need to jointly analyze the network designing from a energy sustainability perspective, as well as from a scalability, operator cost perspective.

Cost inefficient strategies are not likely to incentivize the mobile operators to adopt them and implement on the ground. Hence, various costs borne to the mobile operator like CAPEX and operational expenditure (OPEX) require to be taken into account so as to make the network scalable [35, 46]. At the onset, designing purely green as well as cost profitable frameworks are diverging objectives. This monograph aims to study the tradeoff, associated with designing future networks from cost as well as energy perspective.

1.2.4 Leveraging the Smart Grid Infrastructure

Upgrading the traditional power grid by incorporating renewable energy sources in the demand side (i.e., with the grid users) in additional to the power grid connectivity, along side intelligent communication between the grid operator and the grid users, is termed as Smart grid. Upgrading the traditional power grid towards a ‘smart’ grid involves intelligent control frameworks to coordinate the energy management among the networked nodes or end users using the grid infrastructure itself. By making each (densely deployed) grid user, i.e., BS, a potential energy source, we aim to revise the conventional power generation philosophy in power grids. Furthermore, this direct current (DC) power generation may align well with the upcoming DC power grid concept [47], (instead of classical alternating current (AC) power grid, requiring the DC solar power to be converted to AC for injection into the grid).

The smart grid infrastructure also involves advance sensor and communication infrastructure among the grid users and the grid. The smart grid enables renewable energy powered utility networks to trade energy with the grid. Intelligent renewable energy supply provisioning of the BSs along with leveraging the grid infrastructure can potentially assist in development of self-sustainable, cost-optimal networks [48, 49]. The authors in [50] have designed smart grid connected residential systems, wherein the load is not actually stochastic and can be controlled by the grid user. Cellular communication networks are fundamentally different from residential networks as the cellular communication traffic, i.e., the BS load itself is stochastic and cannot be predetermined.

Additionally, grid connected and ambient powered networks may also face stochasticity in energy harvest due to spatially distributed grid users. The space and time stochastic variation of green energy harvest and BS traffic is referred to as traffic-energy imbalances. This monograph pertains towards proposing novel grid connected and ambient powered communication frameworks for designing energy sustainable, cost profitable, and scalable communication networks; leveraging the grid infrastructure as illustrated through Fig. 1.5. The challenging design aspects associated with the design of grid connected communication networks will be discussed in detail in Chap. 3.

In the upcoming section, a detailed survey of the current state of art frameworks is presented.

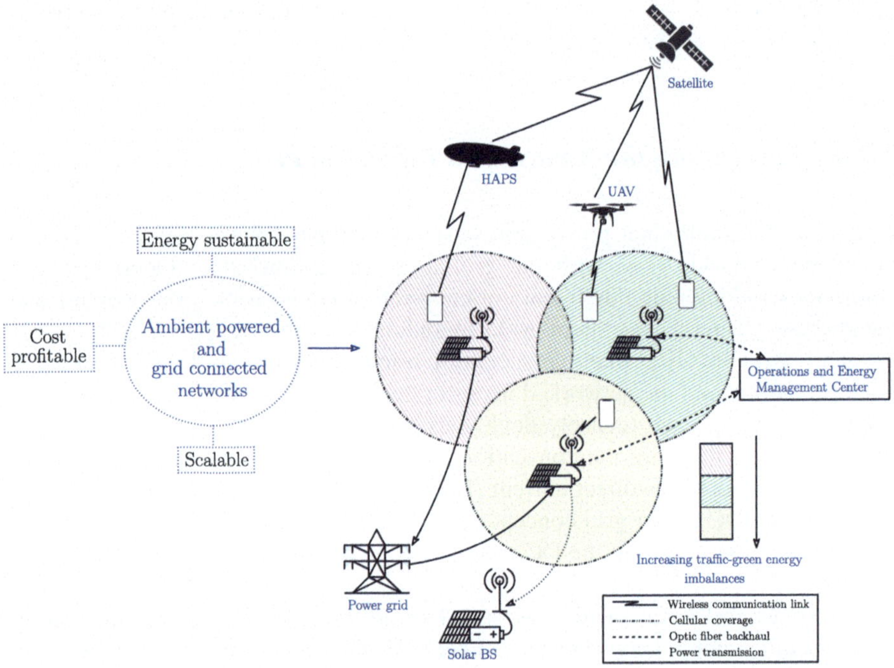

Fig. 1.5 Potential advantages of networked ambient powered intelligent communication systems

1.3 State of the Art

In this section we present a detailed survey of related works around green energy driven networks.

1.3.1 Load Balancing as a Green Communication Strategy

Cellular traffic based load balancing (TLB) or load management strategies have been used traditionally to distribute the system load, aiming to reduce network energy consumption. These strategies include intelligent BS sleeping [51, 52], cell breathing frameworks [53], and intelligent user offloading/ association strategies [48, 49, 54–60]. The motivation behind such analysis is to reduce/offload the energy starved BS by transferring load to other BSs/network access points. Recent frameworks proposed in [51, 52, 54, 55] have considered intelligent BS sleeping strategies to reduce network energy consumption. Intelligently switching the BSs depending on the traffic load and cell zooming techniques [53] have been useful in reducing the network power consumption. The authors in [56, 57, 61–63] studied optimum resource allocation in cellular networks to achieve optimal BS energy consumption.

The authors in [64] presented a comprehensive survey on the modeling and analysis of energy harvesting in grid-powered wireless networks. Optimal resource allocation strategies [61, 62] studied general cellular networks, providing a tractable framework for flexible user association, maximizing the cell capacity, and spectral efficiency. The analysis in [63] presents a comprehensive study on designing cross-layer communication systems with energy constraints. The study in [65] presented an efficient Markov decision process (MDP) based user association strategy in super-WiFi networks, representing a BS state jointly based on the number of users associated and the battery level.

Dual-powered networks also face challenge in dealing with stochastic environments [66]. The randomness in traffic, harvested energy, and channel conditions need to be considered to make the problem more practical. The works presented in [67, 68], and [69] considered the randomness of traffic and harvested energy but did not consider randomness in channel variations. The framework presented in [69] dealt with designing purely solar powered off-grid systems. Even though it considered random traffic, the traffic nature was homogeneous. The user traffic generated for the users associated with the BS can be heterogeneous in nature. The study in [70] dealt with minimizing the grid consumption of mixed traffic green networks, where user satisfaction for each traffic class was represented using utility functions [71].

Spatio-temporal traffic-energy imbalances have been analyzed in recent works [72–75]. The authors in [72] study cellular networks subjected to spatial distributions of traffic load and power consumption, without dealing with green energy or dual-powered supplies. The framework in [73] presents an energy-efficient traffic offloading approach in heterogeneous networks. The framework in [74] analyzes spatial modeling of traffic in cellular networks, while the study in [75] presents a framework for a traffic-aware BS sleeping control strategy to reduce power consumption. While the traffic generated in [74] and [75] are stochastic, they do not consider spatio-temporal traffic variations across the network.

Recent frameworks have also analyzed the possibility of reducing the number of BSs in the network through network coding [37, 38] to achieve energy efficient networks. While this strategy may be useful in homogeneous load scenarios, the cellular networks are generally subjected to space-time varying inhomogeneous loads. In such inhomogeneous traffic scenario, having a limited number of BSs in the network may limit the network performance and reduce the user QoS.

1.3.2 Energy Management Based Grid Connected Network Operation

Energy balancing/management strategies as proposed in [2, 50, 76–78] aim to reduce the network energy consumption by intelligently exploiting the space-time inhomogeneities in BS load and energy harvest, thereby improving temporal green energy utilization.

Green energy sharing among dual-powered BSs has been explored in [76, 77, 79]. In [76], the authors explored an online energy cooperation framework based on MDP to minimize the grid energy acquisition. Energy transfer among the BSs was proposed using a mesh framework of separate power lines between the BSs. The framework presented in [77] improved from the framework in [76] to estimate the optimum mesh connectivity of power lines between individual BSs to facilitate energy exchange between them. Recently, the authors of [79] have also studied energy cooperation among hybrid powered BSs using the Lyapunov optimization framework. The proposed energy sharing framework in [79] is similar to [76], i.e., through a separate mesh network of power lines.

The above studies [76, 77, 79] did not examine the effect of spatio-temporal traffic-energy imbalances on grid energy procurement or operator revenues. Hence, it is very crucial to investigate and understand the effects of these imbalances in a dual-powered cellular network to manifest useful engineering insights. The frameworks in [76, 77, 79] have not evaluated closed form expressions for the optimal amounts of grid-energy procurement and the energy transferable to a BS. Additionally, they did not focus on the specific cost aspects of the network, for instance, these frameworks did not discuss the optimal CAPEX required to design their network. Therefore, system performance evaluation through green energy balancing (GEB) requires to be studied in detail.

The frameworks in [48] and [49] have considered cellular cooperative coverage adjustment based strategy to achieve energy efficient networks. While these frameworks are very effective, there is possibility that due to distributed nature of the BSs, the temporal green energy in the network might not be fully utilized. Additionally, frequent changing antenna power level for cell-breathing frameworks is not very practical and the strategy is computationally heavy for increasing number of BSs.

While energy management strategies are also extremely effective in reducing the grid energy procurement and minimizing the carbon footprint, they are limited by finite green energy storage device capacity and finite antenna power radiation limit [80] due to which a BS cannot serve infinite users associated with it. Hence, due to extremely skewed traffic inhomogeneities, the performance of such energy management strategy degrades resulting in possible reduction in user QoS. Thus, there is a strong motivation to analyze the jointly the effects of load and energy management in a grid connected and communication framework.

1.3.3 Energy Outage Aware Network Frameworks

Traditionally, BS clustering is performed depending on the frequency reuse factor [81]. Modern B5G communication systems generally use unity frequency reuse factor, i.e., full-frequency reuse [82], thus removing the need for frequency allocation based BS clustering.

The existing works pertaining to estimation of energy outage have relied on computationally intensive Markovian frameworks [83, 84]. For instance, the framework in [83], which considers a purely solar provisioned standalone BS, modeled the traffic arrival as well as energy harvest as separate discrete time Markov chains (DTMC) comprising of a large number of states. The framework in [85] proposes a Markov chain based solar energy harvest model, decomposing the energy harvest into numerous states for a residential framework. The authors in [86] propose a Markov chain based strategy to dimension the energy harvesting system in a cellular framework. Additionally, the frameworks in [83, 84] also decompose the battery level into a large number of states, thereby further increasing the system computational complexity.

It is notable that, in a grid connected solar powered system green energy outage needs to be estimated at every time index (depending on the time granularity, it is usually at every hour) [87]. Thus, a Markovian framework comprising of a large number of states, as in [83], is computationally too intensive. Furthermore, since this computation is typically performed in the cloud data-center, it eventually leads to a higher energy consumption [88, 89]. The authors in [67] proposed a framework to optimally dimension an energy efficient standalone solar powered BS.

1.3.4 Joint Traffic-Energy Management Based Networks

Most of the works in the literature have optimized system design either with respect to traffic [48] or green energy [2]. Load management strategies to reduce network energy consumption includes intelligent BS sleeping, cell breathing frameworks, and intelligent user offloading/association strategies [48, 54, 55, 91–100]. The motivation behind such analysis is to reduce/offload the energy starved BS by transferring load to other BSs/network access points. Recent frameworks proposed in [54, 55] have considered intelligent BS sleeping strategies to reduce network energy consumption. The authors in [92, 93, 98] studied optimum resource allocation in cellular networks to achieve optimal BS energy consumption. The authors in [99, 100] proposed cell activation framework to balance energy consumption and QoS satisfaction in wireless networks. While such frameworks are very effective, there is possibility that due to distributed nature of the BSs, the temporal green energy in the network might not be fully utilized. Additionally, frequently altering antenna power level in cell-breathing frameworks may not be very practical and the strategy is computationally heavy with increasing number of BSs.

Table 1.1 Related works with respect to joint traffic energy management based network operation strategy

	Purely renewable energy harvesting without grid connectivity	Grid connected and renewable energy harvesting	Load management	Energy management
[90]		✓		
[54, 55, 57, 59]	✓		✓	
[37, 38]			✓	
[48, 91, 92]		✓	✓	
[2, 50, 76, 77]		✓		✓
JTEB framework		✓	✓	✓

Recent load management based frameworks have also analyzed the possibility of reducing the number of BSs in the network through network coding [37, 38] to achieve energy efficient networks. While this strategy may be useful in homogeneous load scenarios, the cellular networks are generally subjected to space-time varying inhomogeneous loads. In such inhomogeneous traffic scenario, having a limited number of BSs in the network may limit the network performance, thereby reducing the user QoS.

Energy management strategies as proposed in [2, 50, 76, 77] aims to reduce the network energy consumption by intelligently exploiting the space-time inhomogeneities in BS load and energy harvest, thereby improving temporal green energy utilization. While such only sharing of energy (only SE) based frameworks as in [2] are extremely effective in reducing the grid energy procurement and minimizing the carbon footprint, they are limited by finite green energy storage capacity and finite antenna power radiation limit [80] due to which a BS cannot serve infinite users associated with it. Hence, at skewed traffic inhomogeneities, the performance of such energy management strategy degrades resulting in reduction of user QoS.

Beyond the current literature, to fully leverage the effects of space-time varying traffic-energy imbalances, a joint traffic and energy management involving networked cellular coverage adjustment and energy sharing across the cells is considered, in addition to energy trading with the grid. A joint traffic-energy balancing (JTEB) framework may be extremely beneficial in terms of better user QoS, higher service revenue to the operator, as well as green energy sustainability. The lacuna in the literature is illustrated in Table 1.1 and is compared with respect to the JTEB framework, which will be covered in detail as we proceed in the monograph.

1.3.5 Integrated Aerial-Terrestrial Future Networks

Recent research has pertained around integrating the existing terrestrial communication networks with aerial and space networks [101]. Aerial and space communication technologies have evolved around the study of drone based UAVs [102], high altitude platform stations [103], and satellites [104]. Satellite networks used for communication purposes include low earth orbit (LEO), medium earth orbit (MEO), and geostationary earth orbit (GEO) satellites. HAPS are emerging as a very attractive next generation telecommunication technology to integrate with the terrestrial networks as it is averse to energy constraints and latency challenges occurring in drone based UAVs and satellites, respectively [105, 106].

The HAPS is an unmanned aerial vehicle, which operates from the stratosphere (around a height of 20 km). The stratosphere has unique properties which make it very suitable for HAPS operation [107]. There is abundance of sunlight, unrestricted due to clouds, rain, fog, etc., in addition to mild wind speed. These properties are very useful in ensuring the HAPS can be powered with renewable energy sources for its sustenance, and use it for telecommunication and computational operations.

Traditionally, heterogeneous networks involving small cells have been an effective strategy to increase cellular capacity, but they result in extra deployment cost to the operator and are not green. Aerial HAPS mounted radio access networks which can provide coverage up to 500 km, are hence being studied as an attractive solution. Through this monograph, the possibility of offloading excess users to an aerial HAPS mounted BS (hMBS) in the event of high traffic at the terrestrial BS (tBS) or low energy harvest is explored, and the system performance of a green HAPS integrated terrestrial communication architecture is studied.

1.4 Organization

Below we present briefly regarding the structure of this monograph.

- Chapter 2 presents a detailed design and performance evaluation in purely ambient powered, off-grid solar BSs. The chapter first showcases a Markovian framework for probabilistically analyzing the energy outage is such systems. The performance of these off-grid BSs is then evaluated in terms of the optimal CAPEX (i.e., solar provisioning) required to avoid energy outages over a long duration.
- In Chap. 3, we present the grid networked and ambient powered, "dual-powered", communication architecture. We enlist the design constraints and challenges in system analysis, and the associated cost parameters and network operations involved in these networks. Through this chapter, we give a glimpse of the topics which will be covered in the upcoming chapters, namely, the system design objectives, particularly, carbon free green network and operator revenue aware network designing, and further integrated aerial terrestrial networks.
- Through Chap. 4, we first present an energy balancing based in grid networked telecommunication framework. We then showcase the design of an energy sustainable, carbon free network design through an energy balancing based cooperative energy transfer framework. The system analysis includes optimizing

the amount of energy which can be transferred among the networked nodes, a Markovian framework to estimate the energy outages occurring in an energy balancing model, and the optimal cluster size and solar provisioning per BS required to realize a distributed virtual energy bank.

- In Chap. 5, we analyze the grid networked and ambient powered telecommunication framework from an operator perspective. By incorporating revenue awareness, we aim to illustrate the benefic effects of energy balancing and its complementarity to load balancing, especially to aid scalability of green solutions to the industry. From a system perspective, we first analyze and explain the revenue metrics associated with telecom networks. A joint load-energy balancing framework is studied towards optimizing the operator revenue, taking the system constraints into account.
- Through Chap. 6, we aim to study the inherent coupling between the inherent energy and data networks present in the grid networked and ambient powered communication framework. As a case study, the analysis is performed for futuristic HAPS based aerial framework.
- In Chap. 7, we demonstrate a wider perspective of energy balancing in networks, through a residential setting. The emphasis is to highlight that the concept of energy-load balancing is extremely crucial in energy-aware networks including ambient internet of things (IoT), wireless sensor networks, residential networks, or green data-center designing.
- Through Chap. 8, we showcase a glimpse of green AI aided future networks. We start by classifying the green communication strategies as node level, network level, or at a computational level—describing the sustainability challenges in each level. We then outline the current and upcoming AI tools used for green communications. The emphasis is to highlight the tradeoff between energy efficiency and network performance among various AI models, and the need to develop computationally light and efficient AI models.
- Chapter 9 concludes the monograph providing insights to future research directions in this field.

Chapter 2
Ambient Powered Standalone Communication System Design

2.1 Introduction

Traditionally, the BSs have been powered through the power grid [37, 38], which operates through carbon emitting coal based thermal power plants. Perceiving the significant threat of global warming and climate change, it is imperative that the ICT sector also contribute novel strategies to mitigate the carbon footprint. A very effective solution to cut down the carbon footprint, is to power the BSs with renewable energy sources like, solar [40], wind [41], or geothermal energy [42, 43]. In solar abundant countries like India, which is sitting on a cusp of telecommunication revolution, solar energy powered BSs can be a very effective strategy to cut down carbon footprint and thus move toward green communication [44].

Telecommunication companies like Orange (in Europe) and Verizon (in the USA) have been deploying solar powered BSs as a test case, and study the scalability of such systems [108]. Powering BSs with renewable energy sources like solar or wind energy comes with a different set of challenges. These renewable energy sources are prone to climatic changes, i.e., the energy harvesting efficiency is dependent on the weather conditions. For instance, solar energy harvest will be very effective in summer and spring seasons but the harvesting efficiency will be very poor in rainy and fog based seasons. Renewable energy like wind energy cannot be generated in plain land areas and is more prominent in coastal regions. Solar energy in particular cannot be harvested in the night time, hence the system requires to be provisioned to additional battery supplies to supply the system when the harvest is not good, thereby preventing energy outage. Hence, provisioning the BS with renewable energy supplies poses extra CAPEX to the mobile operator [45]. Therefore, studies related with estimating energy outages and deployability of green solutions are important. Below we present key insights which a reader can obtain from this chapter.

L.-C. Wang et al., *Green Energy Driven Integrated Smart Grid and Wireless Networks*, Wireless Networks, https://doi.org/10.1007/978-3-032-20509-4_2

2.1.1 Key Insights from This Chapter

- In this chapter, we present a detailed system level design details towards designing ambient powered standalone BS. The solar BS is off-grid and relies purely on its solar provisioning to sustain.
- Solar BSs are prone to energy outages due to randomness of the solar energy harvest. Hence, we present a Markovian framework to estimate the energy outages.
- The transition probabilities of the battery level (a function of BS load and energy harvest) are derived through first principles. These transition probabilities are used to estimate the energy outage.
- We then present the optimal CAPEX provisioning, i.e., upper limit of CAPEX provisioning, in such off-grid solar BS. The analysis is performed through Parzen window, probability density estimation based technique.

2.2 Ambient Powered Solar Base Station

Let us consider a solar BS, i.e., the BS is solar provisioned (equipped with photo voltaic (PV) panels and storage batteries) as shown in Fig. 2.1. The BS is connected to the core network (termed as backhaul connection) through optic fiber based wired links. The BS sends regular data (traffic/ energy harvest) to the Operations and Energy Management Center (OEMC) implemented at the core network using these backhaul links. Below, we delineate the probabilistic traffic, energy harvest, and BS power consumption. It will be then used to estimate the energy outages through first principles.

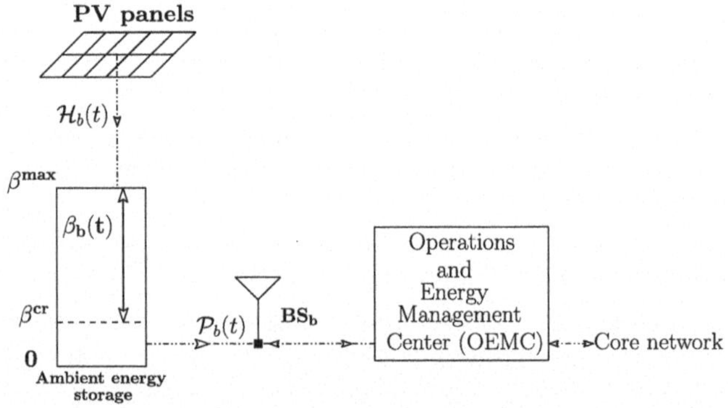

Fig. 2.1 Illustrating a standalone, off-grid, solar BS being powered through solar panels and having its own storage battery

2.2.1 *Traffic and Energy Harvest Modeling*

The traffic arrivals at the solar BS are considered to be distributed as a truncated Gaussian, with hourly varying mean $\mu_L(t)$ and variance σ_L^2. Mathematically, the hourly BS load $L(t) \sim \mathcal{N}(\mu_L(t), \sigma_L^2)$. It is assumed that the users can displace within the network but no new user can enter the network. The hourly energy harvest at the BS $\mathcal{H}(t)$ is modeled as a truncated Gaussian distribution with hourly varying mean $\mu_{\mathcal{H}}(t)$ and variance $\sigma_{\mathcal{H}}^2$. Mathematically, $\mathcal{H}(t) \sim \mathcal{N}(\mu_{\mathcal{H}}(t), \sigma_{\mathcal{H}}^2)$. The hourly energy harvest $\mathcal{H}(t)$ and hourly BS load $L(t)$ are assumed to be uncorrelated and independent to each other.

The hourly power consumption of a BS is computed as [109]

$$\begin{aligned} \mathcal{E}(t) &= N_{trx} P_{\circ} + N_{trx} P_{max} \Delta L(t) \\ &= \Theta_1 + \Theta_2 L(t) \end{aligned} \tag{2.1}$$

where, $\Theta_1 = N_{trx} P_0$ and $\Theta_2 = N_{trx} P_{max} \Delta$ are constants. N_{trx} is the number of transceiver antennas per BS, $P_{\circ}$ is the static power consumption, Δ is the power amplifier efficiency, and P_{max} is the maximum BS downlink power as mandated by Federal Communications Commission (FCC).

Lemma 2.1 *The BS power consumption $\mathcal{E}(t)$, is Gaussian distributed as $\mathcal{E}(t) \sim \mathcal{N}(\Theta_1 + \Theta_2 \mu_L(t), \Theta_2^2 \sigma_L^2)$.*

Proof Since $L(t) \sim \mathcal{N}(\mu_L(t), \sigma_L^2), \mathbb{E}[L^2(t)] = \sigma_L^2 + \mu_L^2(t)$ [110]. Further,

$$\begin{aligned} \mathbb{E}[\mathcal{E}(t)] &= \mathbb{E}[\Theta_1 + \Theta_2 L(t)] = \Theta_1 + \Theta_2 \mu_L(t). \\ \text{Also, } \mathbb{E}[\mathcal{E}^2(t)] &= \Theta_1^2 + \Theta_2^2 \sigma_L^2 + \Theta_2^2 \mu_L^2 + 2\Theta_1 \Theta_2 \mu_L(t). \\ \text{Hence, } var[\mathcal{E}(t)] &= \mathbb{E}[\mathcal{E}^2(t)] - \mathbb{E}[\mathcal{E}(t)]^2 = \Theta_2^2 \sigma_L^2. \end{aligned} \tag{2.2}$$

□

Depending on the hourly energy harvest $\mathcal{H}(t)$ and BS power consumption $\mathcal{E}(t)$, the hourly battery level is computed as follows:

$$\beta'(t) = \beta(t-1) + \underbrace{\mathcal{H}(t) - \mathcal{E}(t)}_{Z(t)} \tag{2.3}$$

$$\text{where, } Z(t) \sim \mathcal{N}(\underbrace{\mu_{\mathcal{H}}(t) - \Theta_1 - \Theta_2 \mu_L(t)}_{\mu_Z(t)}, \underbrace{\sigma_{\mathcal{H}}^2 + \Theta_2^2 \sigma_L^2}_{\sigma_Z^2}). \tag{2.4}$$

Thus, the hourly battery level can be considered as the difference of two random variables $(\mathcal{H}(t) - L(t))$, with $\beta(t-1)$ being a constant scalar at hour t. The battery level, $\beta'(t)$, is distributed as

$$\beta'(t) \sim \mathcal{N}(\beta(t-1) + \mu_Z(t), \sigma_Z^2). \quad (2.5)$$

The battery level is constrained between the upper storage capacity denoted as $\beta^{max} = N_B \times B_{cap}$ and the lower critical threshold denoted as $\beta^{cr} = \delta \times N_B \times B_{cap}$. Here, N_B is the number of batteries equipped with a BS, B_{cap} is the capacity of individual battery, and δ is the depth of discharge. The battery level constrained between β^{cr} and β^{max} is given as

$$\beta(t) = \min\{\max\{\beta'(t), \beta^{cr}\}, \beta^{max}\}. \quad (2.6)$$

In the upcoming section, we present a Markovian framework to estimate energy outages.

2.3 Markovian Characterization of Energy Outages

A standalone solar BS is prone to energy outages due to the stochastic fluctuations of energy harvest and BS load. In this section, we present a Markovian framework, characterizing the battery storage associated with the solar BS and further estimate the hourly energy outage using first principles.

2.3.1 *Markov Modeling and Probabilistic Analysis*

The green energy storage battery level is characterized by a Markov model with three-states, namely, deficit (D), intermediate (I), and surplus (S), as shown in Fig. 2.2. A BS is termed to be in deficit, intermediate, or surplus state if, $\beta_b'(t) < \beta^{cr}$, $\beta^{cr} \leq \beta_b'(t) \leq \beta^{max}$, or $\beta_b'(t) > \beta^{max}$, respectively. The state transition matrix

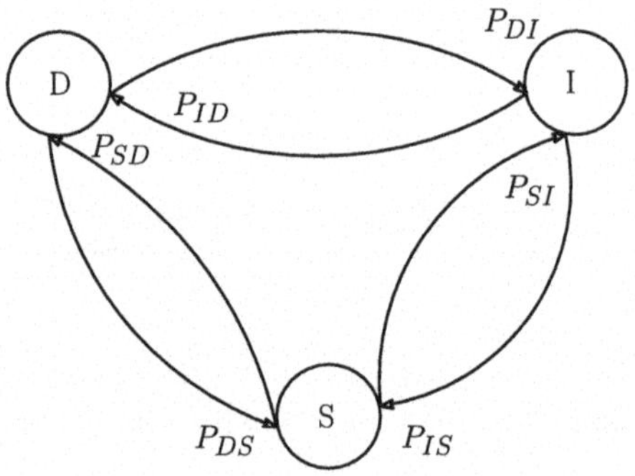

Fig. 2.2 Markovian model of green energy storage in a standalone ambient powered network

corresponding to the three states is represented below as

$$\underline{\mathrm{T}} = \begin{pmatrix} \mathbb{P}_{DD} & \mathbb{P}_{DI} & \mathbb{P}_{DS} \\ \mathbb{P}_{ID} & \mathbb{P}_{II} & \mathbb{P}_{IS} \\ \mathbb{P}_{SD} & \mathbb{P}_{SI} & \mathbb{P}_{SS} \end{pmatrix} \tag{2.7}$$

where $\mathbb{P}_{ij}$ is the probability of transition from state $i \rightarrow j \; \forall \, i, j \in \{D, I, S\}$. Below we derive closed form expressions for each of the transition probability element in the matrix, $\underline{\mathrm{T}}$.

1. *Transition from deficit state to deficit state:*

$$\mathbb{P}_{DD} = \mathbb{P}(\beta_b'(t) \le \beta^{cr} \mid \beta_b(t-1) \le \beta^{cr}) \tag{2.8}$$

$$= \mathbb{P}(\beta_b(t-1) + \mathcal{H}(t) - \mathcal{E}(t) \le \beta^{cr} \mid \beta_b(t-1) \le \beta^{cr}) \tag{2.9}$$

$$= \mathbb{P}(\mathcal{H}(t) - \mathcal{E}(t) \le \beta^{cr} - \beta_b(t-1)) \times \mathbb{P}(\beta_b(t-1) \le \beta^{cr}) \tag{2.10}$$

$$= \mathbb{P}(Z(t) \le \beta^{cr} - \beta_b(t-1)) \times \mathbb{P}(Z(t-1) \le \beta^{cr} - \beta(t-2)) \tag{2.11}$$

$$= \mathbb{P}\left(\frac{Z(t) - \mu_Z(t)}{\sigma_Z} \le \frac{\beta^{cr} - \beta_b(t-1) - \mu_Z(t)}{\sigma_Z}\right) \tag{2.12}$$

$$\times \, \mathbb{P}\left(\frac{Z(t-1) - \mu_Z(t-1)}{\sigma_Z} \le \frac{\beta^{cr} - \beta(t-2) - \mu_Z(t-1)}{\sigma_Z}\right) \tag{2.13}$$

$$= \mathbb{Q}\left(\frac{-\beta^{cr} + \beta_b(t-1) + \mu_Z(t)}{\sigma_Z}\right) \times \mathbb{Q}\left(\frac{-\beta^{cr} + \beta(t-2) + \mu_Z(t-1)}{\sigma_Z}\right). \tag{2.14}$$

2. *Transition from deficit state to surplus state:*

$$\mathbb{P}_{DS} = \mathbb{P}(\beta_b'(t) \ge \beta^{max} \mid \beta_b'(t-1) \le \beta^{cr}) \tag{2.15}$$

$$= \mathbb{P}(Z(t) \ge \beta^{max} - \beta_b(t-1)) \times \mathbb{P}(Z(t-1) \le \beta^{cr} - \beta(t-2)) \tag{2.16}$$

$$= \mathbb{P}\left(\frac{Z(t) - \mu_Z(t)}{\sigma_Z} \ge \frac{\beta^{max} - \beta_b(t-1) - \mu_Z(t)}{\sigma_Z}\right) \times \tag{2.17}$$

$$\mathbb{P}\left(\frac{Z(t-1) - \mu_Z(t-1)}{\sigma_Z} \le \frac{\beta^{cr} - \beta(t-2) - \mu_Z(t-1)}{\sigma_Z}\right) \tag{2.18}$$

$$= \mathbb{Q}\left(\frac{\beta^{max} - \beta_b(t-1) - \mu_Z(t)}{\sigma_Z}\right) \times \mathbb{Q}\left(\frac{-\beta^{cr} + \beta(t-2) + \mu_Z(t-1)}{\sigma_Z}\right). \tag{2.19}$$

3. *Transition from deficit state to intermediate state:*

$$\mathbb{P}_{DI} = \mathbb{P}(\beta^{cr} \leq \beta_b'(t) \leq \beta^{max} \mid \beta_b'(t-1) \leq \beta^{cr}) \tag{2.20}$$

$$= \mathbb{P}(\beta^{cr} - \beta_b(t-1) \leq Z(t) \leq \beta^{max} - \beta_b(t-1)) \times \tag{2.21}$$

$$\mathbb{P}(Z(t-1) \leq \beta^{cr} - \beta(t-2)) \tag{2.22}$$

$$= \mathbb{P}\Big(\frac{\beta^{cr} - \beta_b(t-1) - \mu_Z(t)}{\sigma_Z} \leq \frac{Z(t) - \mu_Z(t)}{\sigma_Z} \leq \frac{\beta^{max} - \beta_b(t-1) - \mu_Z(t)}{\sigma_Z}\Big) \tag{2.23}$$

$$\times \mathbb{P}\left(\frac{Z(t-1) - \mu_Z(t-1)}{\sigma_Z} \leq \frac{\beta^{cr} - \beta(t-2) - \mu_Z(t-1)}{\sigma_Z}\right) \tag{2.24}$$

$$= \left(\mathbb{Q}\left(\frac{\beta^{cr} - \beta_b(t-1) - \mu_Z(t)}{\sigma_Z}\right) - \mathbb{Q}\left(\frac{\beta^{max} - \beta_b(t-1) - \mu_Z(t)}{\sigma_Z}\right)\right) \tag{2.25}$$

$$\times \mathbb{Q}\left(\frac{-\beta^{cr} + \beta(t-2) + \mu_Z(t-1)}{\sigma_Z}\right). \tag{2.26}$$

4. *Transition from intermediate state to deficit state:*

$$\mathbb{P}_{ID} = \mathbb{P}(\beta_b'(t) \leq \beta^{cr} \mid \beta^{cr} \leq \beta_b'(t-1) \leq \beta^{max}) \tag{2.27}$$

$$= \mathbb{Q}\left(\frac{-\beta^{cr} + \beta_b(t-1) + \mu_Z(t)}{\sigma_Z}\right) \times \tag{2.28}$$

$$\left(\mathbb{Q}\left(\frac{\beta^{cr} - \beta(t-2) - \mu_Z(t-1)}{\sigma_Z}\right) - \mathbb{Q}\left(\frac{\beta^{max} - \beta(t-2) - \mu_Z(t-1)}{\sigma_Z}\right)\right). \tag{2.29}$$

5. *Transition from intermediate state to intermediate state:*

$$\mathbb{P}_{II} = \mathbb{P}\left(\beta^{cr} \leq \beta_b'(t) \leq \beta^{max} \mid \beta^{cr} \leq \beta_b'(t-1) \leq \beta^{max}\right) \tag{2.30}$$

$$= \left(\mathbb{Q}\left(\frac{\beta^{cr} - \beta_b(t-1) - \mu_Z(t)}{\sigma_Z}\right) - \mathbb{Q}\left(\frac{\beta^{max} - \beta_b(t-1) - \mu_Z(t)}{\sigma_Z}\right)\right) \times \tag{2.31}$$

$$\left(\mathbb{Q}\left(\frac{\beta^{cr} - \beta(t-2) - \mu_Z(t-1)}{\sigma_Z}\right) - \mathbb{Q}\left(\frac{\beta^{max} - \beta(t-2) - \mu_Z(t-1)}{\sigma_Z}\right)\right). \tag{2.32}$$

6. *Transition from intermediate state to surplus state:*

$$\mathbb{P}_{IS} = \mathbb{P}(\beta_b'(t) \geq \beta^{max} \mid \beta^{cr} \leq \beta_b'(t-1) \leq \beta^{max}) \tag{2.33}$$

$$= \mathbb{Q}\left(\frac{\beta^{max} - \beta_b(t-1) - \mu_Z(t)}{\sigma_Z}\right) \times \tag{2.34}$$

$$\left(\mathbb{Q}\left(\frac{\beta^{cr} - \beta(t-2) - \mu_Z(t-1)}{\sigma_Z}\right)\right.$$

$$\left. - \mathbb{Q}\left(\frac{\beta^{max} - \beta(t-2) - \mu_Z(t-1)}{\sigma_Z}\right)\right). \tag{2.35}$$

7. *Transition from surplus state to deficit state:*

$$\mathbb{P}_{SD} = \mathbb{P}(\beta_b'(t) \leq \beta^{cr} \mid \beta_b'(t-1) \geq \beta^{max}) \tag{2.36}$$

$$= \mathbb{Q}\left(\frac{-\beta^{cr} + \beta_b(t-1) + \mu_Z(t)}{\sigma_Z}\right)$$

$$\times \mathbb{Q}\left(\frac{\beta^{max} - \beta(t-2) - \mu_Z(t-1)}{\sigma_Z}\right). \tag{2.37}$$

8. *Transition from surplus state to intermediate state:*

$$\mathbb{P}_{SI} = \mathbb{P}(\beta^{cr} \leq \beta_b'(t) \leq \beta^{max} \mid \beta_b'(t-1) \geq \beta^{max}) \tag{2.38}$$

$$= \left(\mathbb{Q}\left(\frac{\beta^{cr} - \beta_b(t-1) - \mu_Z(t)}{\sigma_Z}\right) - \right. \tag{2.39}$$

$$\left.\mathbb{Q}\left(\frac{\beta^{max} - \beta_b(t-1) - \mu_Z(t)}{\sigma_Z}\right)\right)$$

$$\times \mathbb{Q}\left(\frac{\beta^{max} - \beta(t-2) - \mu_Z(t-1)}{\sigma_Z}\right). \tag{2.40}$$

9. *Transition from surplus state to surplus state:*

$$\mathbb{P}_{SS} = \mathbb{P}(\beta_b'(t) \geq \beta^{max} \mid \beta_b'(t-1) \geq \beta^{max}) \tag{2.41}$$

$$= \mathbb{Q}\left(\frac{\beta^{max} - \beta_b(t-1) - \mu_Z(t)}{\sigma_Z}\right)$$

$$\times \mathbb{Q}\left(\frac{\beta^{max} - \beta(t-2) - \mu_Z(t-1)}{\sigma_Z}\right). \tag{2.42}$$

Using these transition probabilities, the steady state long-term averaged probabilities are computed from the system of equations given below.

$$\underline{\pi} = \underline{\pi}\underline{\mathrm{T}}, \text{ and } \sum_{i} \pi_i = 1 \tag{2.43}$$

where $\underline{\pi} = \{\pi_D, \pi_I, \pi_S\}$ and $i \in \{D, I, S\}$.

2.3.2 Energy Outage Analysis

In this subsection we will derive the closed form probabilistic expression for consecutive n hour green energy outage in the considered solar BS framework.

Theorem 2.1 *The probability of n consecutive hour green energy outage or blackout in a solar BS system is given as*

$$P_n = \mathbb{P}_{DD}^{n-1}(\mathbb{P}_{DI} + \mathbb{P}_{DS}) \times (\pi_S \mathbb{P}_{SD} + \pi_I \mathbb{P}_{ID}). \tag{2.44}$$

Proof The probability that the Markov model will transit to a deficit state D in the next $n = 1$ consecutive time slots, irrespective of the current state is given by

$$\begin{aligned} P_1 &= \pi_S \mathbb{P}_{SD} \mathbb{P}_{DS} + \pi_S \mathbb{P}_{SD} \mathbb{P}_{DI} + \pi_I \mathbb{P}_{ID} \mathbb{P}_{DI} + \pi_I \mathbb{P}_{ID} \mathbb{P}_{DS} \\ &= \pi_S \mathbb{P}_{SD}(\mathbb{P}_{DS} + \mathbb{P}_{DI}) + \pi_I \mathbb{P}_{ID}(\mathbb{P}_{DI} + \mathbb{P}_{DS}) \\ &= (\mathbb{P}_{DI} + \mathbb{P}_{DS}) \times (\pi_S \mathbb{P}_{SD} + \pi_I \mathbb{P}_{ID}). \end{aligned} \tag{2.45}$$

Similarly, the probability that the Markov model will transit to state D in the next two time slots, irrespective of the current state is given by

$$\begin{aligned} P_2 = {} & \pi_S \mathbb{P}_{SD} \mathbb{P}_{DD} \mathbb{P}_{DS} + \pi_S \mathbb{P}_{SD} \mathbb{P}_{DD} \mathbb{P}_{DI} \\ & + \pi_I \mathbb{P}_{ID} \mathbb{P}_{DD} \mathbb{P}_{DI} + \pi_I \mathbb{P}_{ID} \mathbb{P}_{DD} \mathbb{P}_{DS} \\ = {} & \mathbb{P}_{DD}(\mathbb{P}_{DI} + \mathbb{P}_{DS}) \times (\pi_S \mathbb{P}_{SD} + \pi_I \mathbb{P}_{ID}). \end{aligned} \tag{2.46}$$

By mathematical induction, the probability that the state will remain in deficit state D for n consecutive time slots (hours) is derived as

$$\begin{aligned} P_n = {} & \pi_S \mathbb{P}_{SD} \mathbb{P}_{DD}^{n-1} \mathbb{P}_{DS} + \pi_S \mathbb{P}_{SD} \mathbb{P}_{DD}^{n-1} \mathbb{P}_{DI} \\ & + \pi_I \mathbb{P}_{ID} \mathbb{P}_{DD}^{n-1} \mathbb{P}_{DI} + \pi_I \mathbb{P}_{ID} \mathbb{P}_{DD}^{n-1} \mathbb{P}_{DS} \\ = {} & \mathbb{P}_{DD}^{n-1}(\mathbb{P}_{DI} + \mathbb{P}_{DS}) \times (\pi_S \mathbb{P}_{SD} + \pi_I \mathbb{P}_{ID}). \end{aligned} \tag{2.47}$$

□

In the upcoming section, we present an upper limit to solar provisioning in the ambient solar BS system considered.

2.4 Optimal CAPEX Computation: An Upper Limit to Solar Provisioning

In this section, as a benchmark case, the optimal number of PV panels and storage batteries needed in a standalone solar BS is computed using a data-driven framework, to make the system sustainable and free from energy outages. In the upcoming subsections, we showcase the CAPEX computation using probability density estimation technique.

2.4.1 Probability Density Estimation

CAPEX refers to the cost borne by the mobile operator to install the dual-powered network. It includes the cost involved in equipping the BSs with PV panels and battery storage capacity. In the current off-grid non-networked scenario, the system is mathematically modeled using the traffic and energy harvest profile. Since both traffic load and energy harvest are random in nature, the difference energy between the energy harvested and energy required to serve the BS traffic load at each BS $b \in \mathcal{B}$ is computed. This can be represented as

$$\mathbb{D}_b(t) = \mathcal{H}_b(t) - \mathcal{E}_b(t). \tag{2.48}$$

This difference energy $\mathbb{D}_b(t)$ represents a random variable capturing the stochasticity of both energy harvest and traffic load. The set of difference energies are stored in $\mathbf{D} = \{\mathbb{D}_b(t)\} \; \forall \, b \in \mathcal{B} \; \& \; \forall \, t$.

The difference energy $\mathbb{D}_b(t) \in \mathbb{R}$, i.e., it has support over (∞, ∞) and can take negative values as well as positive values. These two cases are explained below.

- Positive $\mathbb{D}_b(t)$: This event occurs when the temporal energy harvest is greater than the temporal BS load, i.e., $\mathcal{H}_b(t) \geq \mathcal{E}_b(t)$.
- Negative $\mathbb{D}_b(t)$: This event occurs when the temporal energy harvest is lower than the temporal BS load, i.e., $\mathcal{H}_b(t) < \mathcal{E}_b(t)$.

For a stable sustainable system, i.e., a off-grid system without energy outages, the BS should be equipped with sufficient solar provisioning such that, in the event when $\mathbb{D}_b(t)$ is negative, the BS gets the corresponding supply from its battery storage and does not go into outage.

Kernel density estimation strategy with Gaussian kernel is used to estimate the underlying data distribution. This technique is also called as Parzen-Rosenblatt window technique [111]. The probability density function (PDF) estimate using Gaussian kernel Parzen window is represented as a sum of Gaussians and is shown below.

$$\begin{aligned} \hat{f}_{\mathbb{D}}(d_0) &= \frac{\sum_{j=1}^{|\mathbf{D}|} \mathbb{F}(d_0 - d_j, w)}{|\mathbf{D}|} \\ &= \frac{1}{|\mathbf{D}|} \sum_{j=1}^{|\mathbf{D}|} \frac{1}{\left(2\pi w^2\right)^{1/2}} e^{-(d_0 - d_j)^2/2w^2}. \end{aligned} \tag{2.49}$$

Here, $|\mathbf{D}|$ refers to the cardinality of set $\mathbf{D}$, $\mathbb{F}$ refers to the weighting function or kernel function used in the estimation of density points, and w refers to the smoothing parameter or bandwidth of the kernel, which is used to control the window of observation samples for estimating the probability density at a new point.

Mean square error (MSE) is used to measure the performance of the kernel based density estimation and to further compute the optimal Gaussian kernel bandwidth w. It is computed as shown below.

$$\text{MSE}(w) = \mathbb{E}\left[\left(\hat{f}_{\mathbb{D}}(d_0) - f_{\mathbb{D}}(d)\right)^2\right]. \tag{2.50}$$

Here, $f_{\mathbb{D}}(d)$ represents the true density function, while $\hat{f}_{\mathbb{D}}(d_0)$ represents the estimated density function. The optimal kernel bandwidth is chosen by finding the argument that minimizes the above defined MSE function. This is shown as

$$\begin{aligned} w^* &= \arg\min_{w} \text{MSE}(w) \\ &\text{s.t. } w \geq 0. \end{aligned} \tag{2.51}$$

2.4.2 *Optimal Solar Provisioning*

To compute the optimal CAPEX in an off-grid non-networked scenario, the analytical expression of PDF of the difference energy obtained in the previous subsection is utilized. It may be noted that the PDF represented in (2.49) accounts for the daily fluctuations in solar energy harvest as well as the load demand (shown in Fig. 2.4). The PDF effectively captures the temporal stochasticity of difference energy arising due to randomness in harvest energy and load demand profile. The minimum number of solar panels to be provisioned with a BS should satisfy

$\mathbb{E}[\mathbb{D}_b(t)] \geq 0$, or $\mathbb{E}[\mathcal{H}_b(t)] \geq \mathbb{E}[\mathcal{E}_b(t)]$ in a long run. For a system having PV panels of rating $\mathcal{R}$ kW and a lifetime of $\mathcal{L}_{PV}$, the minimum number of PV panels to be provisioned with each BS for a duration of $\mathcal{Y}$ years will be

$$\mathcal{N}_{PV}^{O} \geq \frac{\mathbb{E}[\mathbb{D}]}{\mathcal{B} \times \mathcal{R}} \left(\frac{\mathcal{Y}}{\mathcal{L}_{PV}} \right). \tag{2.52}$$

It is notable that, since the solar provisioning is being statistically estimated based on the annual hourly solar irradiance data obtained from [112], the number of PV panels is estimated such that energy neutrality is maintained. Hence, the expectation of the difference energy should remain positive.

Further, it is observed that the difference energy can be both positive as well as negative. While dimensioning the storage capacity of a BS, it should be ensured that when the energy harvest at any hour is less than the BS load (i.e., difference energy is negative), the BS storage should ensure smooth operation of the BS without energy outage. Thus, the battery storage capacity $\mathbb{C}$ is defined as

$$\mathbb{C} \geq \int_{-\infty}^{0} d_0 \hat{f}_{\mathbb{D}}(d_0) d(d_0). \tag{2.53}$$

Taking the battery depth of discharge δ into consideration, the optimal number of storage batteries $\mathcal{N}_B^O$ of capacity $\mathcal{B}_c$, having a lifetime $\mathcal{L}_B$ to be equipped with each BS is computed as

$$\mathcal{N}_B^O \geq \frac{\mathbb{C}}{(1-\delta) \times \mathcal{B}_c \times \mathcal{B}} \left(\frac{\mathcal{Y}}{\mathcal{L}_B} \right). \tag{2.54}$$

The corresponding CAPEX incurred to the operator is computed as

$$CAPEX^O = \left(C_{PV} \mathcal{N}_{PV}^O + C_B \mathcal{N}_B^O \right) \mathcal{B}, \tag{2.55}$$

where C_{PV} and C_B represent the cost of a unit rated PV panel and battery respectively. The life expectancy of PV panels is generally constant [113], whereas the life expectancy of a battery is dependent on the number of charging-discharging cycles and the depth of discharge set by the operator. Battery lifetime $\mathcal{L}_B$ is computed using the framework in [67], which accounts for the number of charging-discharging cycles, operating temperature T_c (in Celsius), and depth of discharge δ. The dependence of battery cycles C, on T_c and δ is given as

$$C(\delta, T_c) = C(\delta) \times (37.38 \times T_c^{-1.101} - 0.3897). \tag{2.56}$$

Further, the battery lifetime $\mathcal{L}_B$ is computed as

$$\mathcal{L}_B = \frac{1}{\sum_{k=1}^{C(\delta)} \frac{1}{C(\delta, T_c)}}. \tag{2.57}$$

We discuss the key results, observations, and inferences in the upcoming section.

2.5 Results

The data for modeling the cellular traffic has been taken from the data provided in [114]. The energy harvest data used for simulation studies has been obtained from national renewable energy laboratory for New Delhi city, in India [112]. It may be noted that the framework presented is generalized can work on data of any city. Furthermore, we also illustrate the results, when the network is subjected to traffic of varying skewness captured through skewness factor ζ, as in [115, 116]. Computing the optimal CAPEX for extreme skewness, is useful to study the limits of system design and will further assist the operators in designing system for extreme cases. We shall cover the skewed traffic modeling in a greater detail as we cover networked BSs in Chap. 4.

Figure 2.3 shows the variation of consecutive hour energy outage probability in the proposed Markovian framework along with that in the data-driven framework in [83]. We show the variation of the number of consecutive hour k that the DTMC transits in the deficit state, for 72 hours or 3 days. It is observed that the standalone solar BS outage performance modeled through a DTMC closely matches with that in [83]. This also verifies the accuracy of the proposed three-state DTMC to characterize the energy outage.

The histogram plot of the difference energy $\mathbf{D}_b(t)$ presented in (2.48) is captured, and its PDF for various skewed traffic scenarios is shown in Fig. 2.4. It is observed that the PDF takes positive as well as negative values of difference energy, $\mathbb{D}_b(t)$. Clearly, positive values of $\mathbb{D}_b(t)$ indicate the temporal energy harvest greater than the temporal BS load, i.e., $\mathcal{H}_b(t) \geq \mathcal{E}_b(t)$. Similarly, negative values of $\mathbb{D}_b(t)$

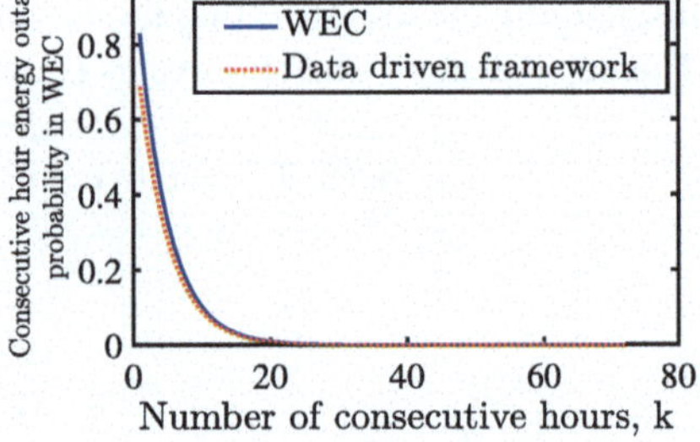

Fig. 2.3 Consecutive hour green energy outage probability in a standalone solar powered BS

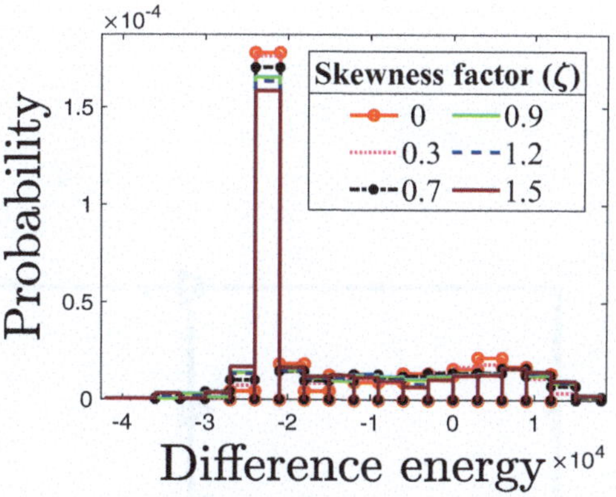

Fig. 2.4 Variation of PDF with ζ

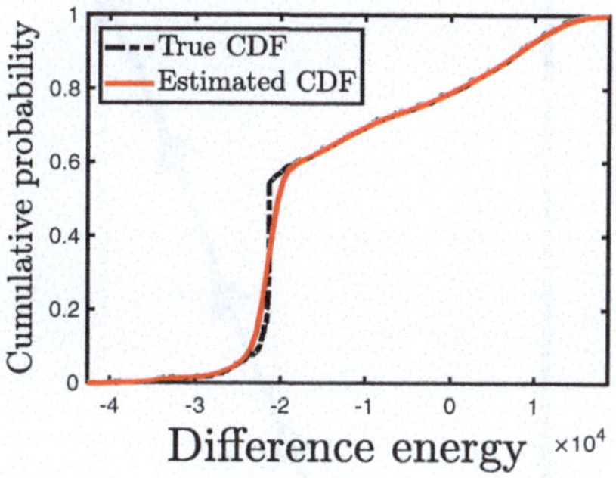

Fig. 2.5 Estimated and the true CDF

Table 2.1 Variation of CAPEX estimation parameters with skewness

ζ	0	0.3	0.7	0.9	1.2	1.5
RMSE	1.2895×10^{-4}	1.2844×10^{-4}	1.2192×10^{-4}	1.661×10^{-4}	1.1408×10^{-4}	1.0956×10^{-4}
w^*	7.0774	7.0916	7.1647	7.2036	7.2588	7.3037

indicate the temporal energy harvest lower than the temporal BS load, i.e., $\mathcal{E}_b(t) > \mathcal{H}_b(t)$. Hence, the support of $\mathbb{D}_b(t) \in \mathbb{R}$.

Through Fig. 2.5 the Parzen window density estimation technique employed to compute the optimal CAPEX in an off-grid non networked scenario is illustrated. Figure 2.5 shows the true and estimated cumulative distribution function (CDF) of the difference energy at $\zeta = 1.5$. It can be observed that the estimated CDF follows the true CDF very closely and results in an root MSE (RMSE) of 1.0956×10^{-4} shown in Table 2.1. The values of optimal kernel bandwidth and the corresponding RMSE values between the estimated and true CDF for various skewed traffic scenarios are shown in Table 2.1.

The optimal solar CAPEX values corresponding to various skewed traffic profiles is illustrated in Fig. 2.6. It is observed that at moderate skewness $\zeta = 0.7$, the CAPEX values increase. Following the increase, the optimal CAPEX values decrease gradually before reaching a stagnation at extreme skewness.

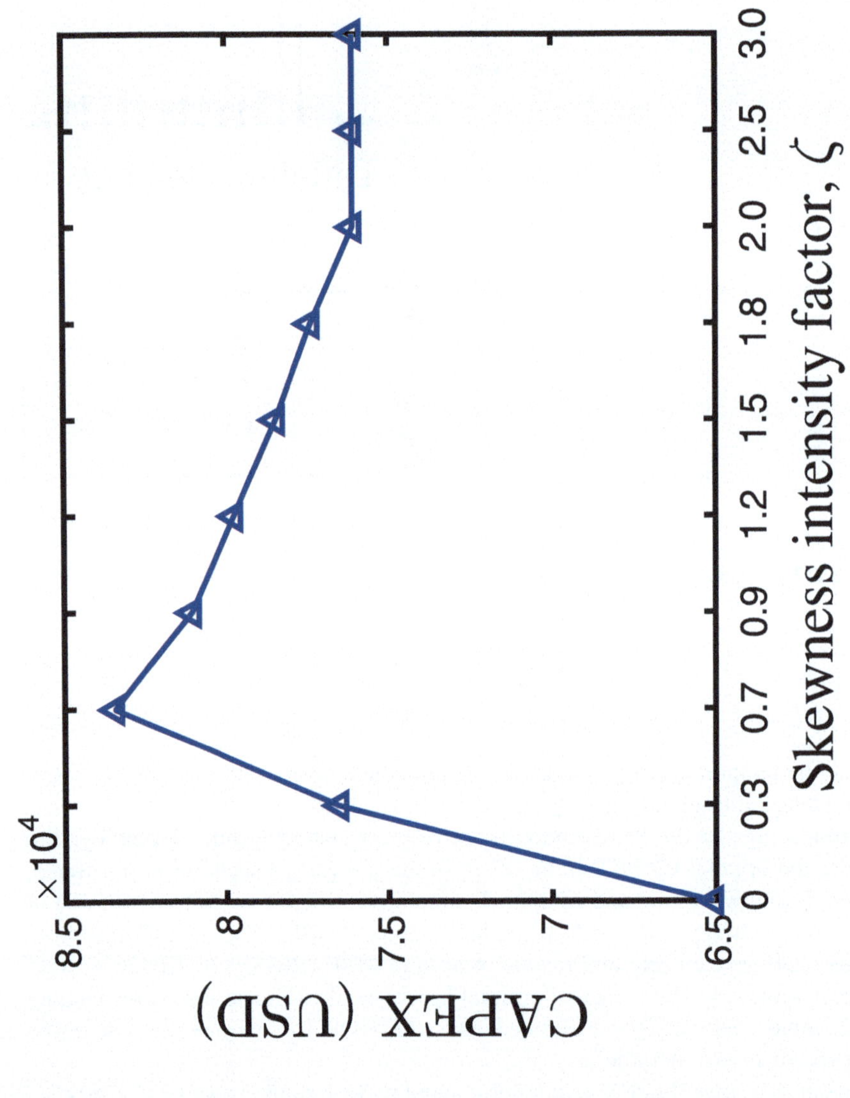

Fig. 2.6 Variation of optimal CAPEX per off-grid BS

2.6 Summary

This chapter has presented a detailed system level design towards designing ambient powered solar BS. Solar BSs are prone to green energy outages, considering the randomness of energy harvest due to climatic conditions. In this chapter, we have shown a Markovian framework to estimate energy outages. Then, we have presented a framework to compute the optimal CAPEX per BS to ensure sustainability. The optimal CAPEX includes the upper limit of solar provisioning per BS. The solar BS is expected to be off-grid and relies purely on ambient solar provisioning to sustain.

Chapter 3
Dual-Powered Networks: A Grid Networked and Ambient Powered Architecture

3.1 Introduction

Green networking solution is a major theme in the upcoming 6G communications [117]. The proliferation of the Internet of Things alongside the evolving 6G communication technology is estimated to significantly increase the quality of user experience. To meet the stringent QoE expectations, which include service guarantee, price paid, as well as mobile device energy consumption, mobile network operators are expected to densify the deployment of BSs, resulting in a potential rise in network energy consumption and thereby carbon footprint [16, 35]. In this chapter, we present a concise in-depth study around the techno-economic aspects involved in designing green future networks. We begin by discussing briefly the recent trends observed in green communications and then entailing the key chapter insights.

3.1.1 Recent Trends in Green Communication

Energy harvesting enabled ambient powered BSs have emerged as carbon free solution. As a test case, mobile operators like Orange, Verizon, and Zong Telecom have considered deploying off-grid ambient powered BSs. While being carbon-free, such purely ambient powered BSs are not scalable due to the high CAPEX [34]. This is mainly because, being purely ambient powered, the BSs require to be significantly over-provisioned by taking into account the temporal climatic changes. Such deployments are also constrained by geographical limitations [44, 118].

Energy-efficient network operation frameworks have considered BS switching involving intelligent powering of the BSs [119]. Despite the promising research being carried out, the network is expected to come under stress under very dynamic spatio-temporally varying network traffic, affecting user QoE. Thus, there is a

L.-C. Wang et al., *Green Energy Driven Integrated Smart Grid and Wireless Networks*, Wireless Networks, https://doi.org/10.1007/978-3-032-20509-4_3

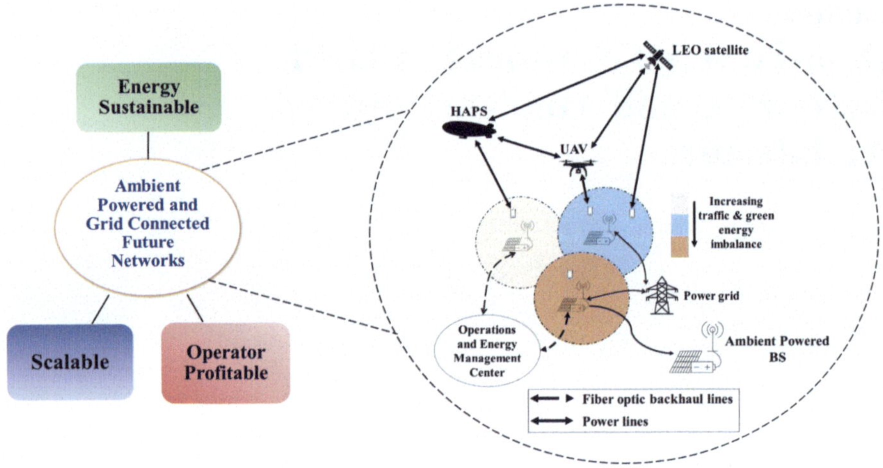

Fig. 3.1 Ambient powered and networked intelligent communication systems: potential advantages and network architecture

pertinent need to design intelligent network operation frameworks catering to user QoE as well as scalability from an industry perspective.

Traditionally, BSs are connected to carbon generating power grid. The power grid connectivity offers scalability if used intelligently. The traditional power grid connectivity is inflexible, i.e., energy can only be purchased by the BS. Designing a system such that the grid users (i.e., BSs) are ambient powered and have energy trading flexibility among themselves as well as with the grid can potentially result in a distributed energy bank, leveraging the grid connectivity [2]. Such smart grid connected and ambient powered BSs have the potential to realize energy self-sustainable and cost-profitable green scalable future networks, as shown in Fig. 3.1. Below we summarize the key contributions in this chapter.

3.1.2 *Key Insights from This Chapter*

1. The chapter presents a comprehensive study for beginners to understand the technical and economic aspects involved in designing green future networks.
2. Through the chapter, we first provide a comprehensive overview of the challenges, requirements, system constraints, inherent cost parameters, and network operations associated with designing an ambient powered and grid connected communication network.
3. The tradeoff between system design objectives is studied especially from a carbon-free *energy-producer mode* and an operator revenue friendly *energy-prosumer mode* of network operation. Further, the study showcases the signifi-

cance of joint load and green energy balancing based network operation strategy over purely load and energy balancing strategies.

4. As an advance, the chapter presents a case study integrating aerial and terrestrial communication technologies towards designing cost profitable green networks. Through a Markovian framework, the study showcases the inherent coupling between the communication and energy networks. Finally, the chapter concludes by highlighting potential upcoming research directions.

3.2 Designing Ambient Powered and Networked Communication Systems

Ambient powered and grid connected networks involve the BSs that are individually equipped with energy harvesting and storage capability. In our current study, as a use-case solar energy is considered to be the ambient energy source with each BS being individually solar provisioned with PV panels and storage batteries in addition to power grid connectivity, as shown in Fig. 3.1. The networked BSs are centrally controlled through an Operations and Energy Management Center, OEMC, designed at the core network. The BSs communicate with the OEMC via fiber optic backhaul lines. The upcoming subsections provide an overview of the system constraints, challenges, and cost parameters to be accounted for while designing ambient powered and grid connected networks as shown in Fig. 3.2.

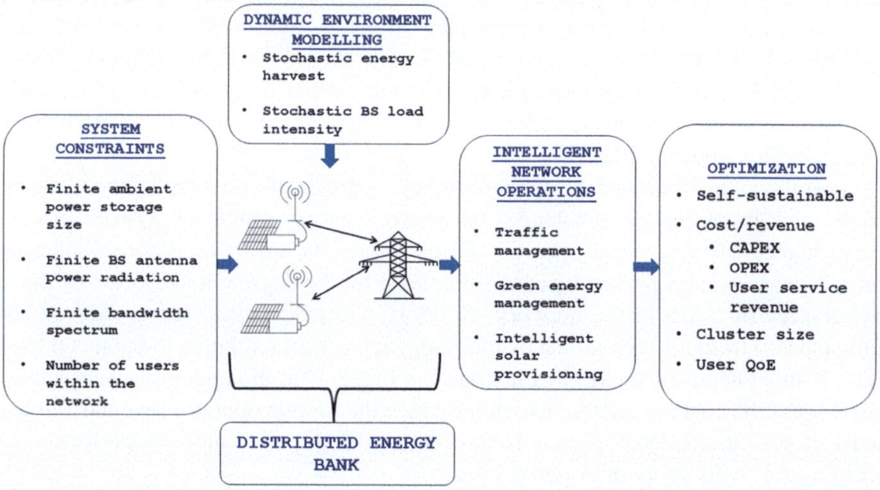

Fig. 3.2 Illustrating the system design overview

3.2.1 Design Constraints and Challenges

In the upcoming 6G communication technology, each BS operates in full frequency reuse mode, i.e., the BS has access to the entire bandwidth spectrum, and thus it is prone to co-channel interference from the other networked BSs [120]. The ambient powered BS is equipped with a finite capacity of PV panels and storage batteries, thus constraining the green energy harvest. Additionally, the BSs have a finite transmit power level as specified by the federal communications commission, FCC, guidelines. [80].

Moreover, ambient powered and grid connected networks experience a dynamic environment involving stochastic renewable energy harvest and BS traffic [121]. This dual stochasticity in green energy harvest and BS traffic varies with space as well as time, resulting in inhomogeneous traffic and green energy imbalances throughout the network. The adverse effects of these imbalances from both user as well as service operator perspectives are listed below.

- Degradation of user QoE
- Reduction of mobile operator revenue
  - Inefficient green energy utilization in the network
  - Higher grid energy consumption

Anecdotally, from the user perspective, there might be network scenario wherein some BSs might become energy-deficient, i.e., the BS has high active user density to be served but does not have sufficient green energy harvest. In such a case, the user network service guarantee gets affected. Further, to meet the user service guarantee the mobile operator will resort to purchasing more energy from the power grid. This will result in not only higher operational costs to the operator, but also contributes to improper green energy management in the network as there might be some other spatially separated BS in the network having surplus green energy in its battery storage which goes unutilized.

Therefore, it is important to capture the effects of the space-time varying traffic and green energy imbalances on network energy efficiency as well as cost performance of the mobile operator. Furthermore, the effect of these imbalances on user and operator performance is influenced by the degree of skewness of these imbalances on a BS in the network. A BS is said to be subjected to skewed (or imbalanced) traffic if it experiences inhomogeneous traffic relative to balanced load (i.e., homogeneous) scenario as illustrated in Fig. 3.1. In the upcoming subsection, we discuss the cost parameters which influence the mobile operator revenue and the network operations which require to be optimized for optimal network performance from user as well as operator perspective.

3.2.2 Cost Parameters and Network Operations

In this subsection we first discuss the cost parameters which influence the design of an ambient powered and grid connected communication network. This will be followed by enlisting the network operations which required to be optimized so as to optimize the network performance.

- *Capital expenditure, CAPEX:* It refers to the deployment cost incurred to the mobile service provider in installing the communication system. It also includes the cost of provisioning the BSs with ambient energy harvesting and storing devices, i.e., cost of solar provisioning each BS.
- *Operational expenditure, OPEX:* It refers to the operational costs incurred to the mobile service provider in daily operations to sustain the network. Prominently, it includes the cost incurred to the BSs in procuring energy from the power grid to avert energy outages. Additionally, if the BSs have the flexibility to trade energy among other networked BSs, then the cost of energy transfer among the BSs using the grid infrastructure is also included in OPEX. This cost is proposed to be borne by the operator towards power grid maintenance.
- *Revenue earned from serving users:* It is the basic source of revenue to the mobile operator, which is paid by the active users in the network in return to the network services. The mobile operator strives to serve as many users possible through the BSs and earn higher revenue.
- *Revenue earned by selling energy:* The mobile operator can earn additional revenue by selling temporal excess green energy in the event of battery overflow or if it is unutilized by any other networked BS at that time.

Generally, the price of purchasing energy from the grid is higher than the price of selling energy to the grid. It is beneficial if a energy starved BS can procure energy from other networked BSs at an energy transfer price before procuring from the power grid. The price of sharing unit energy among the networked BSs is proposed to be higher than the energy selling price but lower than energy purchasing price. This is expected to incentivize the BSs and increase green energy utilization among the BSs rather than selling it to the power grid. It will also deter the BSs from directly procuring energy from the grid at a higher price, thereby reducing the OPEX as well as stemming the carbon footprint.

Below, we enlist the possible BS operations involved in system design. Traditionally, the BSs do not have the flexibility of coverage adjustment or energy sharing to the other BSs. Below, we enumerate the possible BS operations.

- *Traffic load balancing (TLB):* To mitigate the adverse effects of the traffic and green energy imbalances, the networked BSs can resort to cooperative coverage adjustment among themselves through cell breathing frameworks and balance the inhomogeneous space-time varying cellular load among themselves. Load balancing is expected to improve network performance in terms of user QoE [122] as well as operator revenue.

- *Green energy balancing (GEB):* While load balancing is very a promising and energy-efficient strategy, it is not very practical for the mobile operator to vary antenna levels of the BSs frequently in accordance with the temporally changing load. Therefore, green energy balancing among the networked ambient powered BSs through cooperative energy transfer leveraging the power grid infrastructure is expected to be a promising and attractive option [123].
- *Joint traffic load and green energy balancing (JTEB):* The proposed framework involves intelligent load and green energy balancing in an ambient powered and grid connected communication framework. Due to geographically distributed BSs, there is always a possibility of improper green energy utilization in load balancing frameworks. Hence, full green energy potential of the network can be exploited in the JTEB framework through cooperative coverage adjustment as well as cooperative green energy transfer among the networked BSs.
- *Energy trading with the grid:* It involves selling energy back or procuring energy from the power grid.
- *User offloading:* In a multi-tier communication network, it involves intelligently offloading users to another radio access point in the event of a high load or low energy harvest scenario, for network service guarantee as well as improved user QoE.
- *Intelligent solar provisioning:* In a cluster of BSs, depending on the system design objective, intelligent solar provisioning a BS becomes vital. Solar provisioning on a BS not only influences the CAPEX to operator, but also controls the potential green energy flow in the network thereby dictating network performance from an operator's perspective as well as potentially reduced price to be charged to the mobile users for network services.

The upcoming section discusses the key objectives to design a communication system based on the constraints, cost parameters, and network operations presented here.

3.3 System Design Objectives: Analyzing Tradeoffs

An ambient powered and grid connected communication system can be designed from various perspectives. These potentially conflicting performance objectives may not be met together in a single network operation design. Hence, analyzing the tradeoffs involved among various system design objectives is of interest.

3.3.1 Carbon Free Green Network Design

From an environmental perspective, one of the primary motivation is to design carbon free self-sustainable green communication network. In such a network the

BSs do not use power grid energy; the grid infrastructure is used purely for energy selling to the grid or energy sharing among the networked BSs. Hence, the objective of such frameworks is to minimize the grid energy procurement. Since the BSs utilize the grid purely for energy selling/energy transfer purpose, such networks can be termed as an *energy-producer* framework as the BSs act as distributed energy producers to the power grid. The networked BSs through intelligent green energy transfer among themselves, have the potential to act as a self-sustainable distributed energy bank.

While achieving a carbon free network operation, the mobile operator cannot simultaneously maximize the profit, because such green networks generally involve higher CAPEX provisioning per BS so as to offset the OPEX (i.e., grid energy procurement) involved. Another key design objective in ambient powered and grid connected communication networks can be to optimize the BS cluster size required to achieve self-sustainability. Thus, intelligent cooperative coverage adjustment, energy transfer among the networked BSs, and intelligent solar provisioning presents a potentially disruptive solution to realize green scalable future communication networks.

3.3.2 Operator Profitable Network Design

Another perspective of designing an ambient powered and grid connected communication system can be from a mobile operator profit maximization objective. Such a framework will not be green, i.e., carbon free, but will result in intelligent grid energy procurement by the BSs from the power grid. Hence, such a framework will result in BSs being energy producer (i.e., sell energy back to grid), as well as energy consumer (i.e., consumer grid energy). Hence, such networks can be termed as *energy-prosumer* framework. Such energy-prosumer framework will incur much lesser CAPEX provisioning per BS as compared to the carbon free system design discussed in the previous subsection.

3.3.3 Enhancing User QoE

Communication network design can be considered purely from user QoE perspective. While designing the network, it is important for the mobile operator to account for the users' overall QoE, which includes the price paid by user for network service and the user equipment's energy consumption, besides the service guarantee. User service guarantee refers to maintaining signal quality at the user equipment for meeting the rate, throughput, or delay budget requirements.

It may be noted that, guaranteeing user QoE is primary job of the service provider irrespective of whether the communication network is carbon-free or otherwise. Network operations like cooperative coverage adjustment among the networked BSs are greatly beneficial to increase the network service guarantee. This is because coverage adjustment among the spatially distributed BSs assists in better utilization of the network resources (i.e., green energy with each BS) in a traffic-green energy imbalance prone network. For example, connecting a user equipment with appropriate access point in a multi-tier heterogeneous radio access network could aid in battery energy saving in the user equipment. Such green energy enabled network services could also lead to reduced price for network service.

Hence, while designing ambient powered and grid connected networks, it is crucial to account for the physical system constraints and stochastic challenges listed in Section II. Depending on the objectives, the system designing also involves a careful understanding of the cost parameters which can influence the network performance in addition to intelligently selecting the network operation strategies.

3.3.4 Key Performance Studies

Through Fig. 3.3 we illustrate the network performance in terms of user service guarantee and operator profit when the network is operated under various frameworks discussed in Section II-B, namely, without traffic load and energy balancing (wTEB), traffic load balancing (TLB), green energy balancing (GEB), and joint traffic load and energy balancing (JTEB) in an energy-producer setting. To study the system in the long run, the average active users in the network are assumed to be finite such that they can displace within the network. It is inferred that while JLEB and LB provide similar service guarantee but in terms of operator profit, JLEB performs significantly better than LB as the relative network skewness increases. It may be noted that zero relative network skewness corresponds to the homogeneous balanced load scenario, while increasing skewness indicate the increasing traffic inhomogeneity. Thus, while technically network operations may provide similar user QoE, their economic benefits vary from an operator perspective.

Figure 3.4 illustrates the tradeoff in designing a communication system from carbon-free (i.e., energy-producer framework) and operator revenue maximization

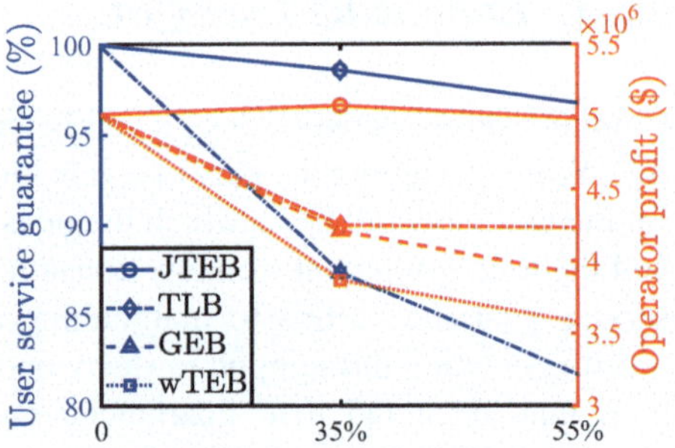

Fig. 3.3 Performance of network operations in terms of user service guarantee and operator profit

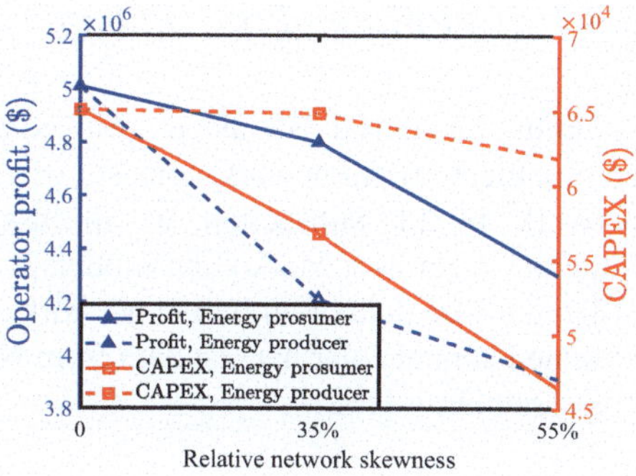

Fig. 3.4 Illustrating CAPEX and operator profit tradeoff between energy-producer and energy-prosumer based framework

(i.e., energy-prosumer framework) perspectives. It is observed that to realize a carbon free communication network, the CAPEX (i.e., solar provisioning per BS) incurred to the operator is higher than the CAPEX incurred in an energy-prosumer framework. It is also observed that the energy-prosumer mode results in significantly higher profit to the mobile operator as compared to the energy-producer mode. Despite the lower profit margins, it may be noted that the operator still gains significant profit through the carbon-free energy-producer mode, thus illustrating the potential benefit of an ambient powered and grid connected communication framework.

3.4 Integrated Aerial Terrestrial Green Networks

In this section, we discuss future communication systems, integrating aerial and terrestrial communication networks towards realizing cost-profitable green communication networks. Subsequently, potential research directions associated with designing more general multi-tier green networks are presented.

Traditional multi-tier terrestrial communication networks involve classification based on macro-micro-pico cell for the deployment, to increase the coverage and capacity. Due to highly dynamic nature of cellular traffic, the exact locations of these smaller cells cannot be pre-determined, resulting in redundant small cell deployments, increasing the network energy consumption and thereby carbon footprint. Aerial/ space technologies like HAPS, UAVs, and LEO satellites have emerged as attractive options to be deployed as an aerial BSs [124]. It is important to study the coupling of the inherent energy and data (i.e., communication) networks. In the upcoming subsections we briefly summarize this inherent coupling. It will be covered in greater detail in Chap. 6.

3.4.1 Coupling of the Data and Energy Networks

A multi tier, aerial BS aided, grid connected, and ambient powered terrestrial communication framework consists of a communication network and an energy network. In this subsection, we discuss these two systems separately and then integrate them as a Markovian model. It may be noted that for this study we have utilized traffic data of an Italian network operator (ref. [25] in [2]). Further we have obtained annual solar harvest data from National Renewable Energy Laboratory for New Delhi city [112].

3.4.1.1 Data Network

In a two tier vertical network comprising of the aerial BS (aBS) and a terrestrial macro BS, tMBS, the user can be associated with either of the BS. The communication network is formed by the downlink between tMBS/ aBS and the active network users. Characterizing the system from a Markovian perspective, users will be offloaded to the aBS only if the tMBS is unable to fulfill the rate guarantee to all the active users. Hence the communication network will transit in either QoE satisfied (Q) state or QoE violation ($\bar{Q}$) state.

3.4.1.2 Energy Network

In an ambient powered and grid connected communication network, the energy network is formed by the state of green energy storage level with the tMBS. The ambient solar energy is harvested through the PV panels equipped with each tMBS and is stored in the local storage battery of the tMBS. The battery supplies energy to the tMBS depending on the temporally active user density and ambient energy harvest at the tMBS. Thus, the tMBS battery level influences the degree of QoE satisfaction.

A tMBS can be characterized as *Deficit (D)*, *Intermediate (I)*, and *Overflow (O)*, depending on the green energy storage level of the BS. The battery storage with the BS has a finite capacity beyond which energy cannot be stored in it. Similarly, the operator decides a critical threshold level below which the battery does not discharge. Hence, the tMBS is termed to be temporally in deficit or overflow state if it does not have adequate green energy storage in its bank to supply to the BS or in the event of battery overflow due to surplus of energy. Else the tMBS is termed to be in intermediate state.

3.4.2 Integrated Markovian Modeling

The data and the energy network models discussed are not independent of each other, rather are coupled as a function of the tMBS energy consumption. Hence, both the Markovian frameworks are dependent on the tMBS power variation which is influenced by the diverse QoE expectations of the temporally active users. Hence, there is a need to analyze the communication and the energy network jointly through an integrated Markovian framework as shown in Fig. 3.5. For designing a green network, the tMBS is assumed to not purchase energy from the grid, rather aim to offload excess users to the aBS as required. Next, we describe the integrated Markov model states and the corresponding network operations.

1. (Q, D): In this state, the tMBS satisfies QoE requirements of the active users. But since battery level is at critical level, some users require to be offloaded to the aBS.
2. (Q, I): The tMBS is sufficient to satisfy QoE of the active users and the battery level is in intermediate state, i.e., above level. This state does not involve either of user offloading or energy selling back to grid.
3. (Q, O): The tMBS is sufficient to satisfy QoE of the active users and battery level is in surplus state. This state only involves energy selling back to the grid and does not involve user offloading to aBS.

Fig. 3.5 Illustrating the Markovian characterization

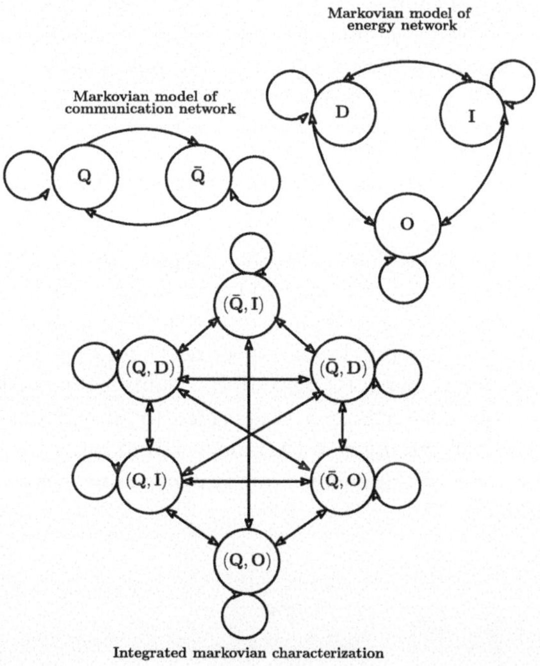

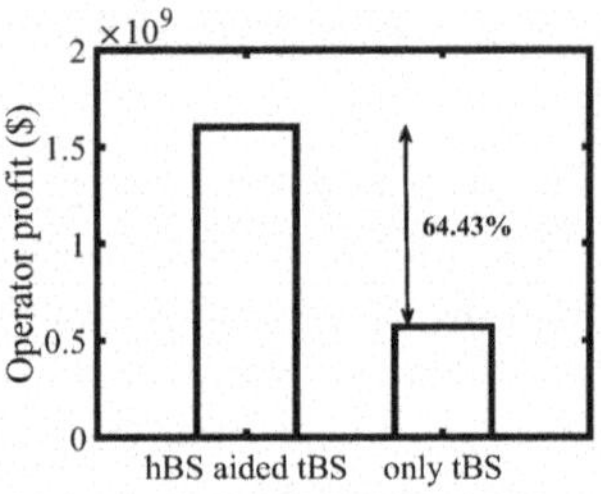

Fig. 3.6 Illustrating operator profit gains in HAPS aided green communication framework

4. $(\bar{Q}, D)$: The tMBS is insufficient to satisfy current QoE needs. Further the battery level is also in deficit state. Hence, this state involves excess user offloading to aBS.
5. $(\bar{Q}, I)$: The tMBS is unable to fulfil the QoE needs, but the battery level is in intermediate state. Thus, this state involves only excess user offloading to aBS.
6. $(\bar{Q}, O)$: The tMBS is insufficient to satisfy current QoE requirements, but battery storage is overflowing. This state involves excess user offloading to aBS and energy selling back to the grid.

As a case study, we consider a two tier system comprising a HAPS-terrestrial network. Performance results shown through Fig. 3.6 illustrate the operator profit in a HAPS aided terrestrial communication framework. It is observed that the aerial offloading based framework is extremely effective in offloading the users in the event of high user density or low energy harvest on the tMBSs, thus improving user QoE in addition to reflecting the operator profit margins. The technical details pertaining to the case study will be presented in detail in Chap. 6 of this monograph.

3.4.3 Future Potential Research Directions

The study presented in this chapter can be potentially extended to generalized ambient powered and grid connected utility networks, for instance residential systems. Future research directions in the domain of ambient powered and grid connected networks are expected to have more diversity, involving space-air-ground integrated networks (SAGIN). Additionally, system performance in presence of dynamic stochasticity is expected to be handled more efficiently by learning and artificial intelligence (AI)-aided optimizations, towards designing self-organizing and robust communication systems. Below we discuss these research directions briefly.

3.4.3.1 Green Residential Networks

The key concept of grid networked ambient powered utility networks presented in this chapter can be potentially extended towards designing green residential systems. Unlike a telecommunication system, the residential load can be potentially controlled by the grid user. The concept of networked energy transfer can be leveraged among the grid connected nodes to reduce the peak hour load demand at the power grid, thus alleviating the power grid demand consumption. Further, temporal grid energy price aware strategies can be potentially formulated by providing monetary benefits at a society level. The concept of micro-grid among residential nodes can be realized towards developing local green energy banks, eventually resulting in lower carbon footprint, in addition to monetary incentives to the grid customers.

3.4.3.2 SAGIN-Aided Green Communication Networks

Traditionally, satellite networks have been used for coverage in rural, remote, and inaccessible areas lacking basic terrestrial infrastructure. SAGIN based vertical heterogeneous networks involving LEO satellites are expected to play a key role in scaling the networking ability in the upcoming 6G networks especially in urban areas [125]. SAGINs are expected to aid in capacity injection through an interplay between satellite, aerial (including UAV and HAPS), and ground communication technologies. Different technologies have their own pros and cons which require to be accounted for while designing the network. For instance, satellites in general have latency constraint due to the much higher path loss, but can provide higher capacity. Despite being energy constrained, UAVs are very attractive due to their shorter altitude, swarming capability, and agility. Mitigating co-channel interference among these vertical communication technology layers is going to be very challenging as aerial, terrestrial BSs, and even LEO satellites are being operated in sub-6 GHz. In addition to spectrum management, handover management is also expected to be a key issue in SAGIN frameworks.

The current SAGIN frameworks are primarily motivated with increasing data rate. With the frequency spectrum of interest expanding towards mmWave and THz frequency bands, aerial and satellite technologies require to be equipped with much larger antenna array structures. The larger antenna arrays and increasing RF chains are expected to contribute to an increased network energy consumption.By exploiting edge intelligence based techniques, data offloading and managing computational resources of the various SAGIN elements based on the user QoE can aid in realizing energy-efficient networks.

3.4.3.3 AI-aided Network Design

For effectively dealing with the stochastic load and service resources, upcoming future networks are expected to be AI driven [126]. Ambient powered networks are heavily influenced by spatio-temporal stochasticity in ambient energy harvest, traffic load, and grid energy pricing. Optimum network design requires accurate prediction and estimation of these parameters in addition to studying their correlation, as they influence the network performance. The statistical spatio-temporal correlation of these parameters can be leveraged to further reduce the estimation complexity. With the upcoming networks becoming more and more edge intelligence driven, it is crucial to analyze the patterns of the past data and communicate only the most semantically significant data so as to increase the network spectral efficiency. Thus, AI driven communication systems can aid in developing robust and computationally efficient green future networks.

3.4.4 Summary

This chapter has presented a concise study to understand the inherent techno-economic aspects involved in designing ambient powered green future networks. A glimpse, comprising of the system design challenges, requirements, system constraints, inherent cost parameters, and network operations have been elucidated. This chapter provides glimpses on the upcoming chapters and may be considered as a primer before going further into them. It gives a bird's eye view of the system design objectives, illustrating the associated tradeoffs. The interplay of data and energy networks in futuristic aerial terrestrial networks is also showcased. The chapter concludes with discussing some potential future research directions for the readers.

Chapter 4
Grid Networked Architectures: An Energy Balancing Study

4.1 Introduction

The networking community has, over the decades, worked on the concept of load balancing. Load balancing has been seen as a very useful concept to improve the network throughput and other key performance metrics like energy-efficiency and spectral efficiency.

With the advance of renewable energy technology, ambient powered networked nodes are becoming increasing popular. Energy harvesting enabled networks can be seen in data centers, ambient IoT, and residential networks, in addition to telecommunication networks. With nodes becoming energy-aware, energy balancing in such networks becomes extremely important.

In this chapter we delve into the networking paradigm of energy balancing. The key theme of the current and the upcoming chapter is to visualize how energy balancing complements load balancing. Energy balancing is not intended to be an alternative to load balancing, but as an key complementing framework, especially when analyzing the economics behind networks.

4.1.1 Key Insights from This Chapter

- In this chapter we showcase an energy balancing based cooperative energy transfer framework in an ambient powered and grid connected communication system.
- Through cooperative energy transfer, we aim to design an energy sustainable, carbon free network, wherein the distributed grid networked BSs transfer energy cooperatively among each other. Optimal amount of energy which can be transferred among the networked nodes is derived and presented.

L.-C. Wang et al., *Green Energy Driven Integrated Smart Grid and Wireless Networks*, Wireless Networks, https://doi.org/10.1007/978-3-032-20509-4_4

- Further, we present a Markovian framework to estimate energy outages in an energy balancing network. Analysis pertaining to optimal cluster size of the BSs required to realize an energy-sustainable cluster is also shown.

4.2 Grid Networked and Ambient Powered Systems

We discuss the system model in this section. A closed area A having U users is considered, wherein the users are modeled as a homogeneous binomial point process of density λ, being covered by $\mathcal{B}$ BSs. At the design and deployment stage, K-Means clustering algorithm is used to compute the optimal BS locations. It is assumed that each BS is individually solar provisioned with $\mathcal{N}_{PV}$ PV panels and η_B storage batteries, in addition to power grid connectivity. Each BS is connected with the OEMC, which is implemented at the core network and communicates with the BSs through optic fiber based backhaul connection. The BSs also have the flexibility to transfer energy among each other through the existing grid infrastructure. Let the BSs and users associated with the BSs be represented as $\mathbf{B} = \{1, 2, \ldots, \mathcal{B}\}$ and $\mathcal{U} = \{U_1, U_2, \ldots, U_b\}$ $\forall$ $b \in \mathcal{B}$, respectively, with the total network users being $U = \sum_b U_b$. Each BS b is assumed to have a coverage area of radius R_b and a downlink transmit power level $\mathcal{P}_b$ with $0 \leq \mathcal{P}_b \leq \mathcal{P}_{max}$. The BSs have the flexibility to modify their coverage areas, share energy among each other, or trade energy with the grid, as motivated in the previous chapter. The upcoming subsections discuss analytical modeling of the BS load intensity and energy harvest profile.

4.2.1 Modeling Cellular Traffic Burstiness

In this subsection, we showcase the modeling of traffic burstiness among cells in a network. This is essential so that we can study the performance of the network under extreme loads and compare it with the homogeneous, balanced load scenario.

The hourly net traffic intensity in area A is denoted as $L(t)$ [114], as shown in Fig. 4.1a. The hourly varying load intensity at a BS b is represented as $L_b(t)$, such that $\sum_{b=1}^{\mathcal{B}} L_b(t) = L(t)$. Additionally, due to dynamic user displacement in the network, the BSs may experience skewed space-time varying traffic which is modeled through a skewness intensity factor ζ. Mathematically $L_b(t) = \Gamma_b(t) \times L(t)$, where $\Gamma_b(t)$ represents the skewness of load intensity subjected on a BS and is given as,

$$\begin{aligned} \Gamma_b(t) &= \frac{e^{\zeta \times b}}{\sum_{b=1}^{\mathcal{B}} e^{\zeta \times b}}, \\ s.t., \sum_{b=1}^{\mathcal{B}} \Gamma_b(t) &= 1. \end{aligned} \tag{4.1}$$

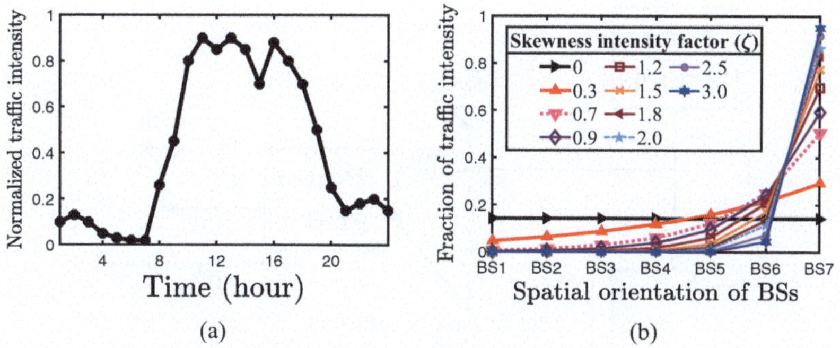

Fig. 4.1 (**a**) Net traffic profile and (**b**) Proposed skewness factor

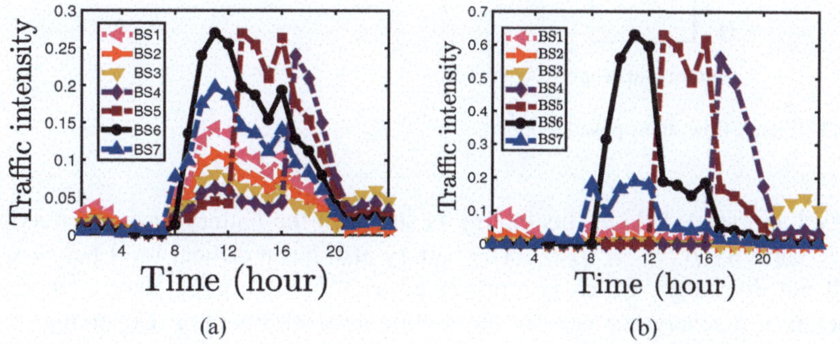

Fig. 4.2 (**a**) Spatio-temporal variation of skewness $\zeta = 0.3$, (**b**) Spatio-temporal variation of skewness $\zeta = 1.2$

Thus, $L_b(t) = (\Gamma_b(t) \times L(t)) = U_b(t)/U$. The skewness factors are illustrated in Fig. 4.1b. This factor weights the burstiness of traffic being experienced in a cell. It may be noted that $\zeta \geq 0$, with $\zeta = 0$ denoting the homogeneous load scenario, whereas $\zeta > 0$ represents increasing load inhomogeneity. The space time variation of user traffic has been illustrated in Fig. 4.2a–b.

4.2.2 Energy Harvest Modeling

Each dual-powered BS is assumed to be equipped with $\mathcal{N}_{PV}$ PV panels and η_B batteries each having a capacity $\mathcal{B}_c$, for energy storage. Let, $\mathcal{H}_b(t)$ be the solar energy harvested at BS b through the equipped PV panels. The annual hourly energy harvest data by a unit rated PV panel is obtained from the National Renewable

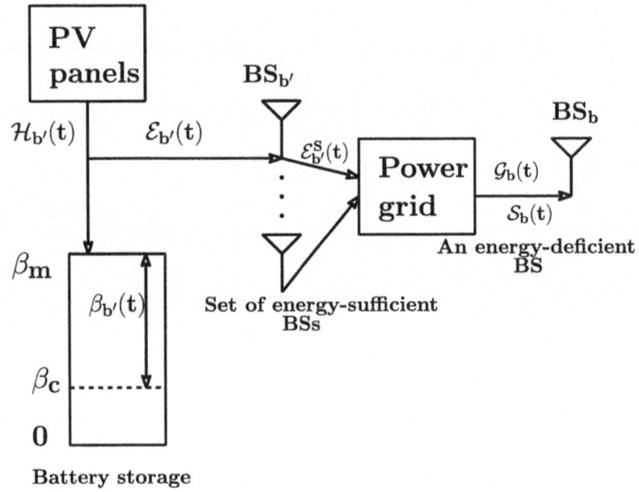

Fig. 4.3 Energy flow in proposed model

Energy Laboratory [112]. This energy is stored in the battery having a maximum storage capacity $\beta^{max} = \eta_B \mathcal{B}_c$. The battery also has a critical level below which it will not discharge, which is represented as $\beta^{cr} = \delta\eta_B\mathcal{B}_c$. Here, δ represents the depth of discharge chosen by the mobile network operator. Depending on the hourly BS energy consumption $\mathcal{E}_b(t)$ and the hourly BS solar harvest $\mathcal{H}_b(t)$, the battery level at each BS b is computed as

$$\begin{aligned} \beta_b'(t) &= \beta_b(t-1) + \mathcal{H}_b(t) - \mathcal{E}_b(t) \\ \beta_b(t) &= \min\left\{\max\left\{\beta_b', \beta^{cr}\right\}, \beta^{max}\right\}. \end{aligned} \tag{4.2}$$

This energy flow in a dual-powered network is represented in Fig. 4.3. Depending on the hourly battery level, a BS is classified as energy deficient, or energy sufficient (indexed by b').

- *Energy deficient BS:* A BS will be *energy deficient* if $\beta_b'(t) < \beta^{cr}$. In this case, the amount of deficient energy required at BS b to avoid energy outage will be $\mathcal{D}(b,t) := \left(\beta^{cr} - \beta_b'(t)\right)$.
- *Energy sufficient BS:* In case a BS is *energy sufficient*, i.e., $\beta_{b'}'(t) > \beta^{cr}$, the amount of sharable energy by BS b' is then defined as $\mathcal{E}^S(b',t) := \left(\beta_{b'}'(t) - \beta^{cr}\right)$.

4.3 Energy Balancing Based Carbon Free Network Design

4.3.1 Cooperative Energy Transfer Based Formulation and Solution

The analysis of a networked dual-powered network first pertains around formulating a grid energy procurement minimization problem. For a given CAPEX, the OEMC classifies the BSs at each hour into two disjoint sets as *energy deficient* or *energy sufficient*, as discussed previously. Let ED out of $\mathcal{B}$ BSs be energy deficient at a given hour, and the remaining $ES = (\mathcal{B} - ED)$ BSs be energy sufficient. For the ED energy deficient BSs, the net hourly deficit energy is

$$\begin{aligned} \mathcal{D}(t) &= \sum_{b=1}^{ED} \mathcal{D}_b(t) \\ &= \sum_{b=1}^{ED} \left(\beta^{cr} - \beta_b'(t)\right). \end{aligned} \tag{4.3}$$

For the remaining ES energy sufficient BSs, the hourly net sharable energy available in the network is

$$\begin{aligned} \mathcal{E}^S(t) &= \sum_{b'=1}^{ES} \mathcal{E}^S(b', t) \\ &= \sum_{b'=1}^{ES} \left(\beta_{b'}(t) - \beta^{cr}\right). \end{aligned} \tag{4.4}$$

In the proposed energy sharing framework, an energy deficient BS can meet its deficit energy requirement in two ways. The energy deficient BS either receives the deficit energy by cooperative energy transfer through the energy sufficient BSs or by energy procurement through the power grid. let $\mathcal{S}_b(t)$ be defined as the portion of deficit energy coming from the ES energy sufficient BSs to an energy deficient BS b and $\mathcal{G}_b(t) = (\mathcal{D}(b, t) - \mathcal{S}_b(t))$ as the remaining portion of deficit energy being procured from the power grid by the deficient BS b.

It is aimed to minimize the grid procurement $\mathcal{G}_b(t) \geq 0$ at each BS, i.e., min $\mathcal{G}_b(t)$. Since, $\mathcal{G}_b(t) = (\mathcal{D}(b, t) - \mathcal{S}_b(t))$, hence min $\mathcal{G}_b(t) \equiv$ max $\mathcal{S}_b(t)$. Ideally for minimizing the grid energy procurement, i.e., for achieving $\mathcal{G}_b(t) = 0$, $\mathcal{S}_b(t) = \mathcal{D}(b, t)$. Hence, from the network perspective,

$$\sum_{b=1}^{ED} \mathcal{S}_b(t) = \min \left\{ \sum_{b'=1}^{ES} \mathcal{E}_{b'}^S(t), \sum_{b=1}^{ED} \mathcal{D}(b, t) \right\} \tag{4.5}$$

and

$$0 \leq \sum_{b=1}^{ED} \mathcal{S}_b(t) \leq \sum_{b=1}^{ED} \mathcal{D}(b,t). \tag{4.6}$$

The corresponding grid energy minimization problem formulation is defined as

$$\begin{aligned}
\mathcal{P}1: \max_{\mathcal{S}_b} & \sum_{b=1}^{ED} \mathcal{S}_b(t) \\
\text{s.t. } 0 \leq & \sum_{b=1}^{ED} \mathcal{S}_b(t) \leq \sum_{b=1}^{ED} \mathcal{D}(b,t) \\
& \sum_{b=1}^{ED} \mathcal{S}_b(t) = \min\left\{\sum_{b'=1}^{ES} \mathcal{E}_{b'}^{S}(t), \sum_{b=1}^{ED} \mathcal{D}(b,t)\right\}.
\end{aligned} \tag{4.7}$$

The problem $\mathcal{P}1$ is observed to be a single variable linear optimization problem with affine and concave constraints. However, it can be observed that the decision variable $\mathcal{S}_b(t)$ vanishes on differentiation. Since it is aimed to compute the optimal amount of energies which can be shared (or procured) to (or by) an energy deficit BS, i.e., the optimum value of $\mathcal{S}_b(t)$ requires to be estimated. Hence, $\mathcal{P}1$ is reformulated as a quadratic convex problem $\mathcal{P}2$, shown below, in order to derive closed-form results and make the problem mathematically tractable.

$$\begin{aligned}
\mathcal{P}2: \max_{\mathcal{S}_b} & \sum_{b=1}^{ED} (\mathcal{S}_b(t))^2 \\
\text{s.t. } C1: & \sum_{b=1}^{ED} \mathcal{S}_b(t) \leq \sum_{b=1}^{ED} \mathcal{D}(b,t) \\
C2: & \sum_{b=1}^{ED} \mathcal{S}_b(t) = \min\left\{\sum_{b'=1}^{ES} \mathcal{E}_{b'}^{S}(t), \sum_{b=1}^{ED} \mathcal{D}(b,t)\right\} \\
C3: & \sum_{b=1}^{ED} \mathcal{S}_b(t) \geq 0.
\end{aligned} \tag{4.8}$$

it is observed that $\mathcal{P}2$ is a purely convex optimization problem. This is because the objective function is convex in nature, constraints C1 and C3 are affine, while constraint C2 is concave in nature.

Lemma 4.1 *For a dual-powered network consisting of $\mathcal{B}$ BSs, the minimum power grid energy procurement required by an energy deficient BS b is*

$$\mathcal{G}_b(t)$$
$$= \begin{cases} 0, \ if \ \sum_{b=1}^{ED} \mathcal{D}(b,t) \leq \sum_{b'=1}^{ES} \mathcal{E}^S(b',t) \\ \mathcal{D}(b,t) \left(\frac{\sum_{b=1}^{ED} \mathcal{D}(b,t) - \sum_{b'=1}^{ES} \mathcal{E}^S(b',t)}{\sum_{b=1}^{ED} \mathcal{D}(b,t)} \right), \ if \ \sum_{b'=1}^{ES} \mathcal{E}^S(b',t) < \sum_{b=1}^{ED} \mathcal{D}(b,t). \end{cases} \tag{4.9}$$

and the maximum energy that can be shared to a energy deficient BS is given as

$$\mathcal{S}_b(t)$$
$$= \begin{cases} \mathcal{D}(b,t), \ if \ \sum_{b=1}^{ED} \mathcal{D}(b,t) \leq \sum_{b'=1}^{ES} \mathcal{E}_{b'}^S(t) \\ \left(\mathcal{D}(b,t) \sum_{b'=1}^{ES} \mathcal{E}_{b'}^S(t) \right) / \sum_{b=1}^{ED} \mathcal{D}(b,t), \ if \ \sum_{b'=1}^{ES} \mathcal{E}_{b'}^S(t) < \sum_{b=1}^{ED} \mathcal{D}(b,t). \end{cases} \tag{4.10}$$

Proof Please refer to Appendix A. □

4.3.2 *Outage Aware Network Design*

This section presents a computationally efficient multi-BS energy balancing based energy cooperation (EC) framework, wherein the networked solar powered BSs have the additional flexibility to share/transfer the green energy stored in individual batteries amongst each other using the existing power grid infrastructure as in Fig. 4.4. The upcoming sections first present the Markovian EC framework, followed by CAPEX planning and operator revenue analysis.

4.3.3 *Markov Modeling*

For a set of $\mathcal{B}$ BSs, each BS in the EC framework is hourly classified as energy-sufficient if $\beta'(t) \geq \beta^{cr}$, or energy-deficient if $\beta'(t) < \beta^{cr}$. The energy-sufficient BSs are proposed to have *transferable green energy* of quanta $\mathcal{E}^S(t) = \beta'(t) - \beta^{cr}$ and the energy-deficient BSs require to be supplied *deficit energy* of quanta $\mathcal{D}(t) = \beta^{cr} - \beta'(t)$.

Let the energy-sufficient BSs at each hour be indexed by b', $s.t.$, $1 \leq b' \leq ES$. Similarly let the energy-deficient BSs be hourly indexed by b, $s.t.$, $1 \leq b \leq ED$. It is notable that a BS can be either energy sufficient or energy deficient, hence the sets of BSs are disjoint to each other, with $ED + ES = \mathcal{B}$.

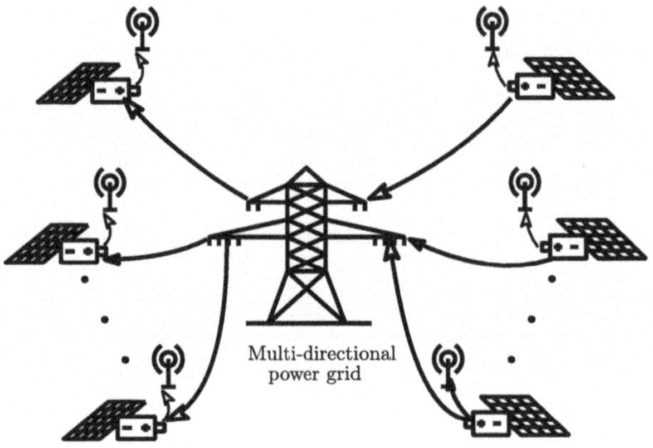

Fig. 4.4 Proposed energy cooperation based communication network

Fig. 4.5 Markovian model of green energy storage in energy cooperation based grid networked solar BS network

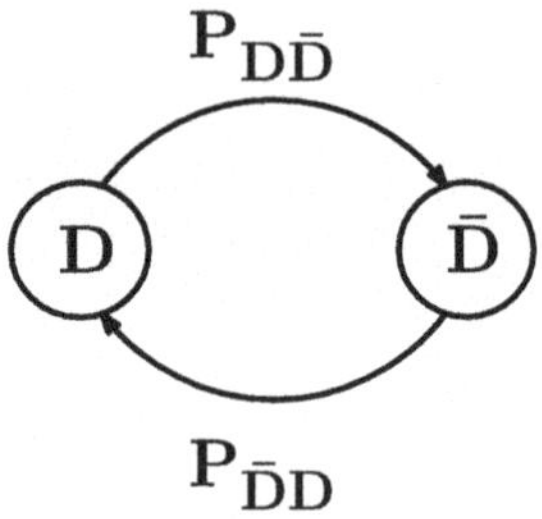

The system of BSs operating in the proposed EC framework is modeled as a DTMC such that at a given hour, the set of $\mathcal{B}$ BSs are in either of the two distinct states, namely, energy deficit (D) and non-deficit ($\overline{D}$) state, as shown in Fig. 4.5. The multi-BS EC system is said to be in state D if the net surplus green energy with the energy-sufficient BSs is less than the net energy deficit required by the energy-deficient BSs. Mathematically,

$$\sum_{b'}^{ES} \mathcal{E}^S(b', t) < \sum_{b}^{ED} \mathcal{D}(b, t). \tag{4.11}$$

The EC system is in state $\overline{D}$ if the net surplus green energy with the networked BSs is greater than the energy deficit experienced by energy-deficient BSs. Mathematically,

$$\sum_{b'}^{ES} \mathcal{E}^S(b', t) \geq \sum_{b}^{ED} \mathcal{D}(b, t). \tag{4.12}$$

Therefore, instead of modeling each BS as a three state DTMC, the entire set of BSs is considered as a two state Markov process, thus considerably decreasing the computational space. The corresponding transition matrix for the EC framework is given as

$$\underline{\mathrm{T}}_{EC} = \begin{pmatrix} \mathbb{P}_{DD} & \mathbb{P}_{D\overline{D}} \\ \mathbb{P}_{\overline{D}D} & \mathbb{P}_{\overline{D}\,\overline{D}} \end{pmatrix}. \tag{4.13}$$

Here, $\mathbb{P}_{ij}$ refers to the probability of transition from state $i \rightarrow j \; \forall \; i, j \in \{D, \overline{D}\}$. The transition probabilities in matrix $\underline{\mathrm{T}}_{EC}$ in (4.13) are defined as follows:

1. *Transition from deficit state to deficit state:*

$$\mathbb{P}_{DD} = \mathbb{P}\left(\sum_{b'}^{ES} \mathcal{E}^S(b', t) < \sum_{b}^{ED} \mathcal{D}(b, t) \,\middle|\, \sum_{b'}^{ES} \mathcal{E}^S(b', t-1) < \sum_{b}^{ED} \mathcal{D}(b, t-1)\right) \tag{4.14}$$

2. *Transition from deficit state to non-deficit state:*

$$\mathbb{P}_{D\overline{D}} = \mathbb{P}\left(\sum_{b'}^{ES} \mathcal{E}^S(b', t) \geq \sum_{b}^{ED} \mathcal{D}(b, t) \,\middle|\, \sum_{b'}^{ES} \mathcal{E}^S(b', t-1) < \sum_{b}^{ED} \mathcal{D}(b, t-1)\right) \tag{4.15}$$

3. *Transition from non-deficit state to deficit state:*

$$\mathbb{P}_{\overline{D}D} = \mathbb{P}\left(\sum_{b'}^{ES} \mathcal{E}^S(b', t) < \sum_{b}^{ED} \mathcal{D}(b, t) \,\middle|\, \sum_{b'}^{ES} \mathcal{E}^S(b', t-1) \geq \sum_{b}^{ED} \mathcal{D}(b, t-1)\right) \tag{4.16}$$

4. *Transition from non-deficit state to non-deficit state:*

$$\mathbb{P}_{\overline{D}\,\overline{D}} = \mathbb{P}\left(\sum_{b'}^{ES} \mathcal{E}^S(b', t) \geq \sum_{b}^{ED} \mathcal{D}(b, t) \,\middle|\, \sum_{b'}^{ES} \mathcal{E}^S(b', t-1) \geq \sum_{b}^{ED} \mathcal{D}(b, t-1)\right) \tag{4.17}$$

In the upcoming subsection, the method used to compute these transition probabilities is explained in detail.

Algorithm 1: Transition probability computation in EC framework

Result: $\mathbb{P}_{DD}, \mathbb{P}_{D\overline{D}}, \mathbb{P}_{\overline{D}D}, \mathbb{P}_{\overline{D}\overline{D}}, \pi_D, \pi_{\overline{D}}$

1 **Input:** $\mathcal{B}, \mathcal{H}(b,t), \rho(b,t), \beta^{cr}, \beta^{max}$

2 **Initialize:** $\mathbb{P}_{DD} = 0, \mathbb{P}_{D\overline{D}} = 0, \mathbb{P}_{\overline{D}D} = 0, \mathbb{P}_{\overline{D}\overline{D}} = 0$

3 Compute $\beta'(b,t)\ \forall b \in \mathcal{B}$ using (5)

4 **for** *b=1 to* $\mathcal{B}$ **do**

5   **for** *t=1 to T* **do**

6   Compute $\mathcal{E}^S(b,t) = \beta'(b,t) - \beta^{cr}, \forall b$

7   Compute $\mathcal{D}(b,t) = \beta^{cr} - \beta'(b,t), \forall b$

8   **if** *condition in* (4.14) **then**

9             $\mathbb{P}_{DD} \leftarrow \mathbb{P}_{DD} + 1$;

10   **else if** *condition in* (4.15) **then**

11             $\mathbb{P}_{D\overline{D}} \leftarrow \mathbb{P}_{D\overline{D}} + 1$;

12   **else if** *Condition in* (4.16) **then**

13             $\mathbb{P}_{\overline{D}D} \leftarrow \mathbb{P}_{\overline{D}D} + 1$;

14   **else**

15             $\mathbb{P}_{\overline{D}\overline{D}} \leftarrow \mathbb{P}_{\overline{D}\overline{D}} + 1$;

16

17   **end**

18 **end**

19 $\mathbb{P}_{DD} = \mathbb{P}_{DD}/T, \mathbb{P}_{D\overline{D}} = \mathbb{P}_{D\overline{D}}/T, \mathbb{P}_{\overline{D}D} = \mathbb{P}_{\overline{D}D}/T, \mathbb{P}_{\overline{D}\overline{D}} = \mathbb{P}_{\overline{D}\overline{D}}/T$

4.3.4 Probabilistic and Complexity Analysis

For the multi-BS EC model, the transition are computed as shown in Algorithm 1. The protocol takes as input the number of BSs, the energy harvest at each BS, the load at each BS depending on the degree of skewness [2], and the critical and upper limit of green energy storage. The transition probabilities are initially assigned with value zero (in Step 2). The amount of energy which can be shared by a energy sufficient BS is computed in Step 6, while the amount of deficit energy with an energy deficient BS is computed in Step 7. Steps 8–17 illustrate the computation of transition probabilities for the proposed EC model.

It can be inferred from Algorithm 1 that the complexity of the proposed EC model is $\mathbb{O}(\mathcal{B} \times T \times S^2)$, where, $\mathcal{B}, T, S$ represent the number of BSs, length of time horizon, and the number of states in the Markovian framework, respectively. Thus, the computation complexity is directly proportional to the number of states and the number of BSs.

The steady state probabilities can be computed for the EC model using the system of equations given below.

$$\begin{aligned} \underline{\pi}_{EC} &= \underline{\pi}_{EC}\underline{\mathrm{T}}_{EC} \\ \sum \pi_i &= 1. \end{aligned} \tag{4.18}$$

Here, $\underline{\pi}_{EC} = \{\pi_D, \pi_{\overline{D}}\}$ & $i \in \{D, \overline{D}\}$. In the upcoming subsection, the probability of consecutive hour energy outage for a EC model is derived.

Theorem 4.1 *The probability of k consecutive hour energy outage or blackout for the EC model is given as*

$$P_k = \pi_D \mathbb{P}_{\overline{D}D} \mathbb{P}_{DD}^{k-1} \mathbb{P}_{D\overline{D}} \tag{4.19}$$

Proof Kindly refer Appendix B. □

4.3.5 Optimal EC Cluster Size and CAPEX Planning for Sustainability

The current subsection evolves around computing the optimal BS cluster size and the CAPEX required in a multi-BS cluster to achieve grid energy independence or self-sustainability. It may be noted that an energy-deficient BS can meet its energy deficit either by cooperative energy transfer from the energy-sufficient BSs ($\mathcal{S}(b, t)$) or procure energy from the power grid ($\mathcal{G}(b, t)$). Hence,

$$\mathcal{D}(b, t) = \mathcal{S}(b, t) + \mathcal{G}(b, t). \tag{4.20}$$

The net energy which can be cooperatively shared among the grid connected solar powered BSs is

$$\begin{aligned} &\sum_{b}^{ED} \mathcal{S}(b, t) = \min\left(\sum_{b}^{ED} \mathcal{D}(b, t), \sum_{b'}^{ES} \mathcal{E}^S(b', t)\right), \\ &\text{with, } \sum_{b}^{ED} \mathcal{G}(b, t) = \sum_{b}^{ED} \mathcal{D}(b, t) - \sum_{b}^{ED} \mathcal{S}(b, t). \end{aligned} \tag{4.21}$$

To eliminate grid energy procurement, the net transferable green energy among the energy-sufficient BSs should be at least equal to the net deficit energy in the network. This condition can be mathematically written as,

$$\sum_{t=1}^{T} \sum_{b=1}^{ED} \mathcal{D}(b, t) \leq \sum_{t=1}^{T} \sum_{b'=1}^{ES} \mathcal{E}^S(b', t). \tag{4.22}$$

From an energy sustainable viewpoint, the joint optimal BS cluster and CAPEX computation optimization is shown below.

$$
\begin{aligned}
P_1 :\ & \min_{\mathcal{B}, N_{PV}, N_B} \ \mathcal{B} \times (C_B N_B / \mathcal{L}_B + C_{PV} N_{PV} / \mathcal{L}_{PV}) \\
s.t.,\ C1 :\ & N_B^O \geq N_B \geq 0 \in \mathbb{Z}^+, N_{PV}^O \geq N_{PV} \geq 0 \in \mathbb{Z}^+ \\
C2 :\ & \mathcal{B} = ED + ES \geq 2 \in \mathbb{Z}^+ \\
C3 :\ & \pi_D = 0.
\end{aligned}
\tag{4.23}
$$

The optimization problem P_1 jointly optimizes the number of networked BSs $\mathcal{B}$ along with the solar provisioning required at each BS, for energy sustainability. Since all the variables being optimized in P_1 are integers ($\mathbb{Z}^+$), P_1 is an integer programming problem. $C1$ constrains the number of number of PV panels and storage batteries, while $C2$ captures the minimum required BSs for EC. Constraint $C3$ represents the fulfillment of (4.22), i.e., grid energy independence achieved by the cluster of $\mathcal{B}$ BSs, with the steady state probability of deficit becoming zero. P_1 is solved using exhaustive search in MATLAB.

It may be noted that each BS is equipped with a finite number of storage batteries. Further, a BS is mandated to radiate up to P_{max} power as per the FCC guidelines, thereby limiting its coverage [109]. Thus the proposed energy transfer based EC framework is physically limited by two key physical system parameters, namely, battery storage capacity and BS power radiation limit. The complexity analysis of the proposed framework is shown in the upcoming subsection.

4.3.5.1 Complexity Analysis

Referring to the CAPEX optimization problem in (4.23), Constraints $C1$ and $C2$ have a complexity of $\mathbb{O}(\mathcal{B} \times N_B^O \times N_{PV}^O)$, where N_B^O and N_{PV}^O denote the optimal bounds on solar provisioning derived in Chap. 2. Constraint $C3$ has a complexity of $\mathbb{O}(\mathcal{B} \times T \times S^2)$ as shown in Sect. 4.3.4 . Thus the overall complexity of P_1 is $\mathbb{O}(\mathcal{B} \times (N_B^O \times N_{PV}^O + T \times S^2))$, that is the complexity is linear with respect to the number of BSs, time, and solar provisioning per BS, but is quadratic with respect to the number of Markovian states in the framework.

4.3.6 Modes of Network Operation

An energy balancing based network can be operated in two possible modes of operation. The first strategy is devised from the environmental perspective of reducing the carbon footprint and is termed as "energy producer mode". The second strategy is devised from the operator's perspective of maximizing revenue and is termed as "energy prosumer mode". Below we explain these modes of network operation.

4.3.6.1 Energy Producer Mode

The carbon free energy producer mode of network operation involves the operator running the network such that the networked BSs act as a distributed energy producer to the power grid. In this mode of network operation, it is assumed that the BSs, without the flexibility to procure energy from the grid, can cooperatively share energy amongst each other and/or sell energy back to the power grid. Thus, in this mode while the BSs are connected to the power grid, they are powered only through green solar energy harvested in their battery storage.

4.3.6.2 Energy Prosumer Mode

The energy prosumer mode relates to the scenario wherein the networked BSs being dual-powered act as both, energy producers to the grid (i.e., sell energy to the grid) as well as energy consumers (i.e., procure energy from the grid). In this mode of network operation, the BSs in addition to energy procurement and selling, also have the flexibility to share energy amongst each other cooperatively.

Optimal CAPEX varies for both the proposed modes of network operation. For the energy producer mode, optimal CAPEX refers to the CAPEX at which the cellular network can operate without any grid energy procurement. For the energy prosumer mode, optimal CAPEX refers to the CAPEX at which the operator can earn maximum revenue.

4.4 Results

4.4.1 Reduction in Grid Procurement

In this subsection, the reduction in grid energy procurement achieved through the proposed prosumer mode of cooperative energy sharing framework is illustrated. It may be noted that the proposed producer mode is already a carbon free mode and hence has not been shown.

In Fig. 4.6a, as a baseline comparison, the grid energy procurement in the energy-balancing based prosumer mode in [2] is compared with a dual-powered without energy cooperation (WEC) based framework considered in [1]. Additionally, to show the magnitude of reduction in carbon emissions, the grid energy procurement in a fully grid-powered system [61] is also illustrated. It can be observed that the prosumer mode of network operation results in a significantly lower grid energy procurement as compared to [1] and [61].

Figure 4.6b shows the average reduction in grid energy procurement with increasing traffic inhomogeneity over [1]. It is observed that the homogeneous traffic condition ($\zeta = 0$) does not involve any reduction in grid energy. This is

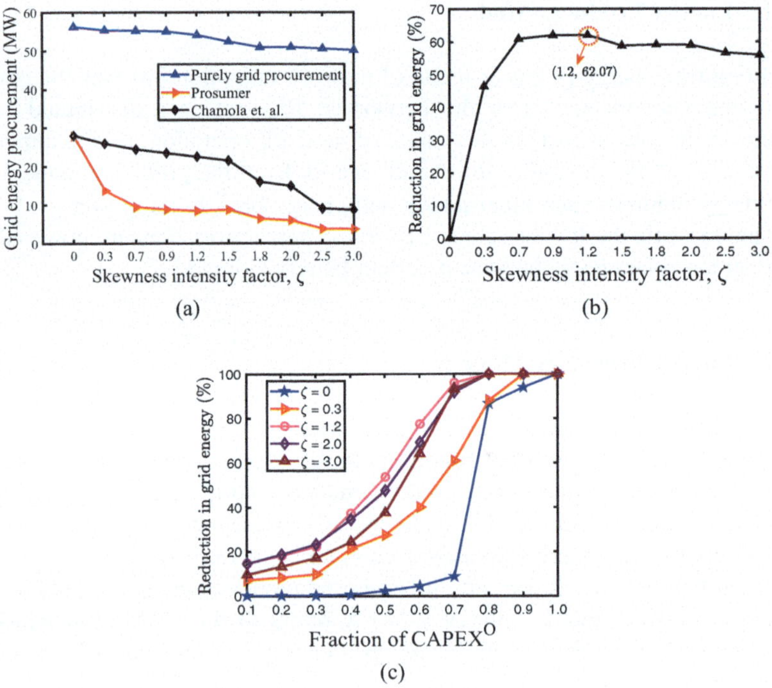

Fig. 4.6 (**a**) Comparison of grid energy procurement in various modes of network operation, (**b**) Average reduction in grid procurement with ζ over [1], (**c**) Variation of grid reduction with CAPEX

essentially due to balanced loads on all the BSs in the network and same energy harvest assumed on all the BSs. Further it is observed that with increasing traffic skewness, the system is able to reduce up to 62.07% of its grid energy procurement (at $\zeta = 1.2$). The average grid energy procurement reduction is observed to decrease marginally after $\zeta = 1.2$, saturating at around 56%.

Remark 4.1 Cooperative energy transfer results in significant reduction of grid energy procurement over a dual-powered WEC based framework.

Through Fig. 4.6c, the variation in grid energy procurement with increasing CAPEX for various skewed traffic scenarios is illustrated. It may be noted that $CAPEX^O$ represents the optimal CAPEX required in the off-grid non-networked scenario, as discussed in Chap. 2. It is observed that when the system is subjected to homogeneous traffic, then the system is able to be carbon free only at full CAPEX (i.e., at $CAPEX^O$). The reduction in grid energy procurement at $\zeta = 0$ is attributed to increasing $CAPEX^O$, which eventually results in lowering the grid energy procurement. On the contrary, the proposed cooperative energy sharing framework is advantageous with increasing traffic inhomogeneity. It is observed that subjecting the system to skewed traffic ($\zeta > 0$) results in achieving self sustainable networks at a significantly lower CAPEX.

Table 4.1 Comparison of number of Markovian states

State of the art	Number of Markovian states, S
TCOMM 2016 [83]	For single BS: 3 (energy harvest) × 2 (BS load) × 36 (Battery level states) = 216
TSE 2015 [84]	For single BS: 4 (energy harvest) × 2 (BS load) × 24 (hour of day) = 192
Grid connected standalone solar BS (Chap. 2)	For single BS: 3
Energy cooperation based dual powered networks	For the BS cluster: 2

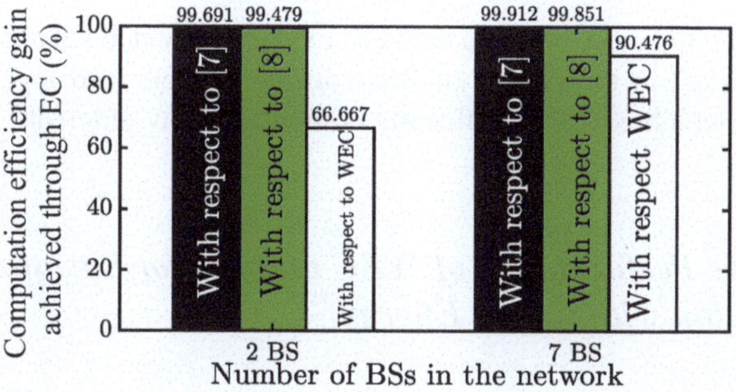

Fig. 4.7 Computation efficiency gain in EC framework over state-of-art

Remark 4.2 The proposed energy producer mode is expected to achieve towards development of self sustainable dual-powered cellular networks, at much lower CAPEX.

Table 4.1 captures the number of Markov states in our proposed approaches in contrast with the number of states in the state-of-art [83, 84]. The Markov modeling in [84] was aimed at computing the green energy outage occurring in standalone solar powered BS, wherein a day was characterized as good/bad depending on the solar energy harvest and a day was further quantized into four time windows. The study in [83] extended the work in [84] by additionally characterizing the BS load along with solar harvest, and battery storage was modeled by multiple Markov states. The total number of Markov states required in [84] was 192 for battery energy state characterization for a standalone BS, whereas in [83] it required 324 Markov states. In contrast, a standalone, purely solar powered without EC (WEC) framework (presented in Chap. 2), characterizes a standalone solar BS by a three-state Markov model, whereas the multi-BS EC system (with ≥ 2 BSs) characterizes the entire cluster of BSs into two states, to compute the desired green energy outage.

In Fig. 4.7, the computational efficiency gains achieved at the OEMC are depicted, for green energy outage estimation and resource provisioning in the proposed EC framework. It is noted that when the BSs do not have flexibility to share energy amongst each other, the proposed WEC modeling is computationally efficient in estimating green energy outages as compared to the state of art. In a 2-BS network, the proposed WEC modeling framework is about 30% more computationally efficient compared to the approaches in [83] and [84]. As the number of BSs in the network is increased to 7, the gain with the WEC model is about 10%.

Figure 4.7 further reveals that, for energy outage estimation in a cluster of BSs with EC, modeling the energy state of the distributed energy bank is significantly computationally efficient compared to modeling the individual BS battery state. The computational efficiency gain over the techniques in [83] and [84] is over 99%. The gain with respect to the proposed WEC model is also significant, respectively about 60% and 90% for a 2-BS cluster and a 7-BS cluster.

Remark 4.3 In an EC system with a network of grid connected solar powered BSs, green energy storage in the cluster of BSs can be characterized together using a two-state model, which is also significantly computationally efficient.

4.4.2 Relative Performance of WEC and EC for Sustainable Operation with Green Energy

To compare and observe the benefits of the distributed energy bank based EC strategy over the WEC mode, a cluster of 7 BSs that are solar provisioned and grid connected are subjected to varying levels of traffic skewness and operate the network through the two strategies. Specifically, a cluster of 7 BSs has been considered as the cluster size is lower than the optimal cluster size required to attain sustainability (at any skewness level), as will be discussed through Fig. 4.8. Hence, the proposed EC framework will be prone to energy outages at different skewness levels. Thus, the efficacy of the proposed EC framework to reduce the average energy outage probability can be verified with respect to the baseline WEC framework at a non-energy sustainable cluster size, as shown through Fig. 4.8a–b. Through Fig. 4.8a, the variation of the long term average steady state energy outage probability is illustrated, when the system operates through both the strategies under the influence of skewed inhomogeneous traffic. It can be observed that the WEC model consistently attains a significantly higher steady state outage probability as compared to the multi-BS EC model. The percentage reduction in outage probability achieved with the EC model over the WEC model is also depicted in Fig. 4.8a. As a general trend, it is inferred that with increasing skewed nature of traffic, the percentage reduction in energy outage with the EC framework increases significantly over the WEC model, indicating the advantage of cooperative energy transfer.

Figure 4.8b shows the variation of consecutive hour energy outage in an under-provisioned solar powered multi-BS EC system under different traffic skewness. It is observed that in general, the probability for all skewed traffic levels tend to zero as k increases. Additionally it is inferred that at a moderate skewed traffic, $\zeta = 0.3$,

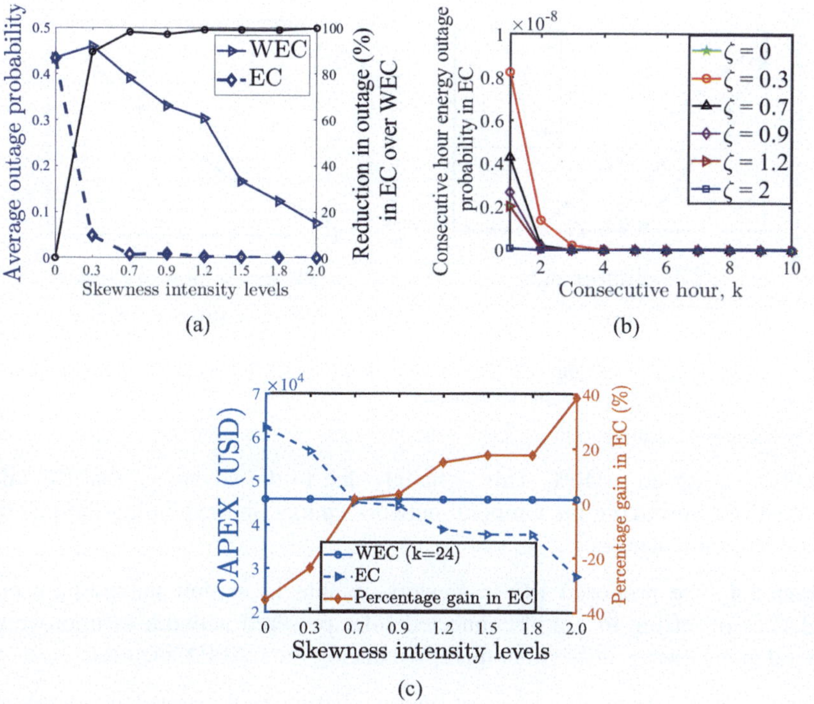

Fig. 4.8 (**a**) Average outage probability and reduction in green energy outage in a cluster of 7 BSs operating through the WEC and EC frameworks, with $\mathcal{N}_{PV} = 5, \mathcal{N}_B = 5$ per BS; (**b**) consecutive hour green energy outage probability for various degrees of traffic skewness in a 7 BS EC framework with $\mathcal{N}_{PV} = 5, \mathcal{N}_B = 5$ per BS; (**c**) CAPEX required to achieve green energy sustainability in WEC and EC frameworks

the EC system experiences highest probability of consecutive hour green energy outage. As the system is subjected to more skewed traffic ($\zeta = 0.7, 0.9, 1.2, 2$)) the probability is relatively lesser. It is also inferred that when the system is subjected to homogeneous traffic ($\zeta = 0$), then the initial probability of consecutive hour energy outage is lower than moderately skewed traffic, but higher than extremely skewed traffic ($\zeta = 2$) . The nature of the plots in Fig. 4.8 for all ζ are inferred to be exponentially decreasing and converging in nature. Hence, P_k proposed through (4.19) in Theorem 2 can be inferred as convex and converging in nature. Since (4.19) is based on the transition and steady-state probabilities computed through Algorithm 1, hence, transition probabilities in $\underline{\mathrm{T}}_{EC}$ and π_D converge over long run, thus guaranteeing convergence of the proposed framework.

Figure 4.8c illustrates the variation of CAPEX required to achieve sustainability or grid energy independence in the WEC and EC systems, when the network is subjected to skewed traffic of varying degrees. It is observed that at a homogeneous traffic scenario, the EC obtains a negative gain over the WEC, but as the degree of skewness increases, the EC performs significantly better than the WEC, obtaining

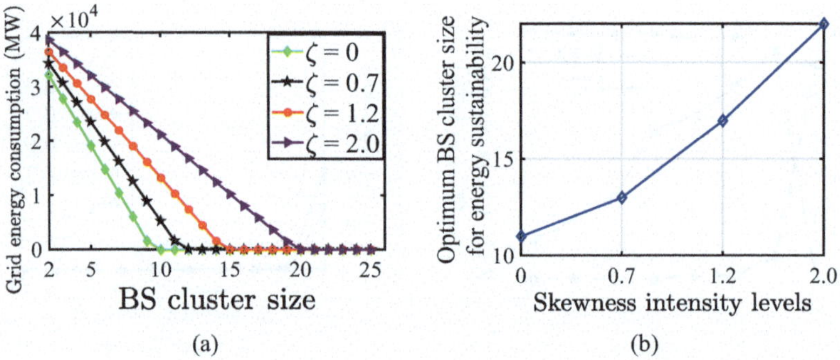

Fig. 4.9 Green energy sustainability in EC framework: (**a**) required BS cluster size; (**b**) optimum BS cluster size as a function of traffic skewness

a CAPEX saving up to 40%. This is mainly due to the advantage that EC takes over WEC by improving the temporal network energy and exploiting the inherent traffic-energy imbalances.

Remark 4.4 The proposed EC framework is able to exploit the traffic-energy imbalances occurring in a grid connected solar powered network to improve the temporal green energy utilization, thereby reducing the CAPEX incurred.

Through Fig. 4.9a–b, the effect of energy sharing is illustrated in a clustered multi-BS EC system, towards attaining energy sustainability when the network. It may be noted that a grid connected solar powered BS requires to procure energy from the power grid, if its energy harvest is unable to meet the load requirements. The proposed EC model provides flexibility to an energy deficit BS to meet its deficit via the unused green energy present with energy-sufficient BSs, thus exploiting the space-time imbalance in green energy harvest and BS traffic. Once the networked BSs, through collaborative energy sharing achieve energy sustainability, then grid energy procurement is not required any further and the grid infrastructure is used purely for sharing energy amongst each other and selling the excess energy to the power grid.

It can be observed in Fig. 4.9a that in general the grid energy consumption reduces in the network as the BS cluster size keeps increasing. Further, it is inferred that higher levels of traffic inhomogeneity tend to require larger BS cluster size. It is observed from Fig. 4.9b that at a balanced/homogeneous load ($\zeta = 0$), around 10 BSs collaborate with each other through the proposed EC framework to achieve grid energy independence, whereas a much increased traffic inhomogeneity ($\zeta = 2.0$) requires a cooperative effort from around 22 BSs to achieve energy sustainability.

4.5 Summary

This chapter has presented an energy balancing based cooperative energy transfer framework in an ambient powered grid connected communication system. Through cooperative energy transfer, it has been aimed to realize energy sustainable, carbon free, green network. The system analysis has included optimal amounts of energy to be shared or purchased from the grid by the networked nodes. A Markovian framework has also been shown to estimate energy outages in an energy balancing based network. The energy balancing solution is expected to be attractive to industry, so as to realize cost efficient green networks.

Chapter 5
Operator Revenue Aware Architectures: A Joint Load and Energy Balancing Paradigm

5.1 Introduction

When we talk about "sustainability", it needs to include "economic sustainability" in addition to "energy sustainability".

In the previous chapter, we demonstrated how energy balancing among grid connected solar BSs (Fig. 5.1a) aids in designing energy-sustainable, carbon free networks. In this chapter, we illustrate how energy balancing (Fig. 5.1b) helps in complementing load balancing (Fig. 5.1c). We observe the effect of load balancing, energy balancing, and joint load and energy balancing (Fig. 5.1d), on key network performance indicators of QoS and economic modeling. It is inferred that while load balancing and joint load and energy balancing perform similarly when in comes to user QoS, but the economically, joint load and energy balancing provides significant gains over load balancing. Thus, illustrating the importance of energy balancing in a framework, in addition to load balancing.

This chapter also demonstrates the significance of proper resource (green energy and power) management in a network, thereby a need for joint load and energy balancing.

5.1.1 Key Chapter Insights

- In this chapter, we design operator revenue aware networks and illustrate as to how energy balancing helps in complementing load-balancing, especially with respect to scaling green solutions to the industry.
- We analyze and explain the revenue metrics associated with designing communication networks. Through a joint load and energy balancing framework, we then showcase the parameters associated with maximizing the operator revenue.
- The tradeoffs associated in designing a carbon free network and an operator revenue greedy framework is also discussed.

L.-C. Wang et al., *Green Energy Driven Integrated Smart Grid and Wireless Networks*, Wireless Networks, https://doi.org/10.1007/978-3-032-20509-4_5

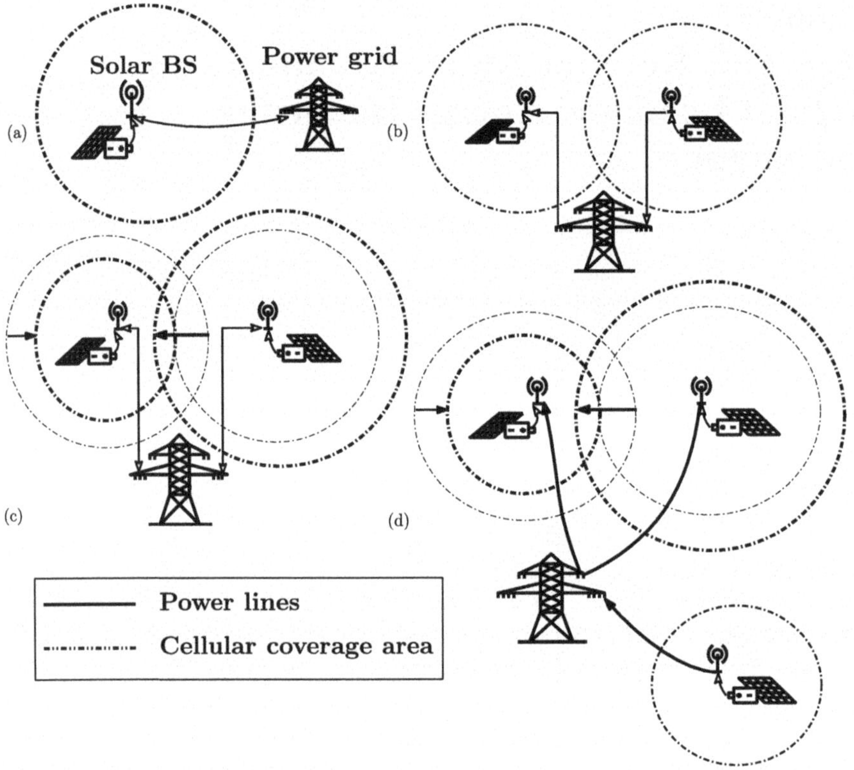

Fig. 5.1 Illustration of (**a**) isolated grid-connected and renewable energy powered BS (as a use case, solar power has been considered in this work), (**b**) BS load aware energy balancing framework, (**c**) BS green energy aware load balancing framework, (**d**) Joint load-green energy balancing framework

5.2 Operator Revenue Aware Network Design

We carry forward the system model in the previous chapter, wherein we have a network of grid connected solar BSs. First, we analyze the key revenue parameters associated with system design.

5.2.1 Key Revenue Metrics

The net profit/loss to a mobile service operator depends on five cost parameters: revenue earned by serving users (R_{serv}), revenue earned by selling energy back to the power grid (R_{sell}), cost incurred to service operator in energy sharing (C_{share}), cost incurred to service operator in purchasing energy from the grid (C_{buy}), and

CAPEX incurred by the service operator in solar provisioning the BSs ($CAPEX$). It may be noted that C_{buy} is included for a general analytical framework, though in a networked and grid-connected BS with green energy aware CAPEX the BSs will not procure energy from the grid. Also, the energy sharing cost through the power grid connectivity is borne by the mobile service operator towards power grid infrastructure maintenance.

The cost parameters are formally defined below.

- *CAPEX:* CAPEX involves the cost to be borne by the mobile network operator in solar provisioning the BSs. This is computed by taking into account the life expectancy of the PV panels and storage batteries. Considering a PV panel having a life expectancy $\mathcal{L}_{PV}$ and a storage battery having a life expectancy $\mathcal{L}_B$, CAPEX for $\mathcal{Y}$ years is computed as

$$CAPEX = \mathcal{B} \times \left(\frac{\mathcal{N}_{PV} C_{PV}}{L_{PV}} + \frac{\mathcal{N}_B C_B}{L_B} \right). \tag{5.1}$$

- *Operational expenditure (OPEX):* In a networked operation scenario, OPEX relates to the cost incurred by the operator in procuring energy from the power grid and in cooperatively sharing energy amongst the BSs using the power grid infrastructure. They are defined below.

1. *Energy sharing cost* (C_{share}): This cost can also be perceived as a grid maintenance cost borne by the operator. The optimum energy to be shared by an energy sufficient BS (4.10) is utilized to compute C_{share} as defined below.

$$C_{share} = \sum_{t=1}^{24} C^{sh} \times \sum_{b=1}^{ED} \mathcal{S}_b(t). \tag{5.2}$$

Here, C^{sh} refers to the cost of sharing unit energy by the sufficient BSs to deficit BSs, using the power grid infrastructure. This expenditure is proposed to be borne by the deficit BSs towards power grid maintenance. It is proposed that C^{sh} be less than the cost of unit grid energy procurement and the cost of selling unit energy back to the power grid, in order to incentivize the power grid operator for energy sharing rather than selling energy.

2. *Grid energy procurement cost* (C_{buy}): It refers to the cost incurred by the network operator in procuring energy from the power grid. It is computed as

$$C_{buy} = \sum_{t=1}^{24} C^{buy} \times \sum_{b=1}^{ED} \mathcal{G}_b(t). \tag{5.3}$$

Here, C^{buy} refers to the cost of procuring unit energy from the power grid. Further, the net OPEX incurred to the operator is given as

$$OPEX = C_{buy} + C_{share}. \tag{5.4}$$

- *Revenue earned by selling energy* ($\mathcal{R}_{sell}$): This is defined as the revenue earned by mobile operator through selling energy back to the power grid. The amount of energy that the operator can sell and earn revenue is given as

$$\sum_{b'=1}^{ES} \mathcal{E}_{b'}^{sell}(t) = \sum_{b'=1}^{ES} \mathcal{E}^S(b', t) - \sum_{b=1}^{ED} \mathcal{D}(b, t), \text{ iff } \sum_{b'=1}^{ES} \mathcal{E}^S(b', t) > \sum_{b=1}^{ED} \mathcal{D}(b, t). \tag{5.5}$$

Also it is observed that,

$$\sum_{b=1}^{ED} \mathcal{S}_b(t) = \min\left\{\sum_{b'=1}^{ES} \mathcal{E}^S(b', t), \sum_{b=1}^{ED} \mathcal{D}(b, t)\right\}.$$

Thus, for $\sum_{b'=1}^{ES} \mathcal{E}^S(b', t) > \sum_{b=1}^{ED} \mathcal{D}(b, t)$, the amount of energy to be sold can be expressed as,

$$\sum_{b'=1}^{ES} \mathcal{E}_{b'}^{sell}(t) = \sum_{b'=1}^{ES} \mathcal{E}^S(b', t) - \sum_{b=1}^{ED} \mathcal{S}_b(t). \tag{5.6}$$

Hence, $\mathcal{R}_{sell}$ is mathematically computed as

$$\begin{aligned} \mathcal{R}_{sell} &= \sum_{t=1}^{24} C^{sell} \times \sum_{b'=1}^{ES} \mathcal{E}_{b'}^{sell}(t) \\ &= \sum_{t=1}^{24} C^{sell} \left(\sum_{b'=1}^{ES} \mathcal{E}^S(b', t) - \sum_{b=1}^{ED} \mathcal{S}_b(t)\right), \end{aligned} \tag{5.7}$$

with C^{sell} referring to the cost of selling unit energy back to the power grid.
- *Revenue earned by serving users* ($\mathcal{R}_{serv}$): It refers to the revenue earned by the mobile operator through serving users. It is computed as

$$\mathcal{R}_{serv} = \sum_{b=1}^{\mathcal{B}} \sum_{t=1}^{24} C^{ser} \times \mathcal{U}_b(t). \tag{5.8}$$

Here, C^{ser} [127] refers to the per-day revenue earned by the mobile operator by serving a user.

5.3 Network Operator Revenue Maximization Analysis

The unconstrained operator profit maximization problem is,

$$\max P = R_{serve} + R_{sell} - C_{share} - C_{buy} - CAPEX. \tag{5.9}$$

For a given CAPEX, the net profit can be expressed as

$$P' = \sum_{t=1}^{T}\sum_{b=1}^{\mathcal{B}} C^{Ser} U_b(t) + \sum_{t=1}^{T}\sum_{b=1}^{\mathcal{B}} C^{sell}\mathcal{E}_b^{sell}(t) - \sum_{t=1}^{T}\sum_{b=1}^{\mathcal{B}} C^{sh}\mathcal{S}_b(t) - \sum_{t=1}^{T}\sum_{b=1}^{\mathcal{B}} C^{buy}\mathcal{G}_b(t). \tag{5.10}$$

Here, it may be noted that P' is not a function of time, rather it is a function of CAPEX provisioning. The variables, C^{sell}, C^{sh}, and C^{buy}, respectively denote the price of selling, sharing, or purchasing unit energy to (through or from) the grid. C^{ser} denotes the daily revenue earned by serving the users.

As discussed in Sect. 4.2.2, depending on the BS green energy storage, temporally, a BS can either be energy-deficient or energy-sufficient. It is proposed that an energy-deficit BS can meet the deficit energy by two ways: either the BS can utilize the sharable green energy in the network through cooperative energy transfer by the energy-sufficient BSs $\mathcal{S}(t)$ or procure the deficit energy directly from the power grid. Thus from a network perspective, $\mathcal{D}(t) = \mathcal{S}(t) + \mathcal{G}(t)$, where $\mathcal{G}(t)$ denotes the energy to be procured by energy-deficient BSs from the power grid.

A deficit BS is incentivized to meet its existing deficit from energy-sufficient BSs and only then procure from the power grid. Similarly, the energy-sufficient BSs after meeting the requirements of the deficient-BSs can sell the surplus sharable green energy back to the power grid. To achieve this, the price of energy transfer from the networked BSs is set to be higher than energy selling price but lower than energy purchasing price, i.e., $C^{sell} < C^{sh} < C^{b}$. In this scenario, the energy-deficient BSs have a cheaper alternative of meeting the deficit energy through energy transfer by the networked energy-sufficient BSs, rather than directly purchasing energy from the grid. Similarly, the energy sufficient BSs also have an incentive to earn a higher price for their sharable green energy rather than selling it cheaply to the grid.

Two specific scenarios may happen at a time instant t depending on the amounts of net deficit or sharable energy, i.e., $\mathcal{D}(t)$ or $\mathcal{E}^S(t)$. These two scenarios are detailed below.

$$\text{If } \mathcal{E}^S(t) \geq \mathcal{D}(t): \begin{cases} \mathcal{G}(t) = 0 \\ E^{sell}(t) = \mathcal{E}^S(t) - \mathcal{D}(t) \\ \mathcal{S}(t) = \mathcal{D}(t). \end{cases} \tag{5.11}$$

$$\text{If } \mathcal{E}^S(t) < \mathcal{D}(t): \begin{cases} \mathcal{G}(t) = \mathcal{D}(t) - \mathcal{E}^S(t) \\ E^{sell}(t) = 0 \\ \mathcal{S}(t) = \mathcal{E}^S(t). \end{cases} \tag{5.12}$$

In the above equations, $E^{sell}(t)$ represents the temporal energy which can be sold by energy-sufficient BSs to the power grid and $\mathcal{S}(t)$ represents the amount the temporal green energy which is shared among the networked BSs. From (5.11) and (5.12), it is inferred that,

$$\begin{aligned} \mathcal{S}(t) &= \min\{\mathcal{D}(t), \mathcal{E}^S(t)\}, \\ \text{and } E^{sell}(t) &= \min\{\mathcal{E}^S(t) - \mathcal{D}(t), 0\}. \end{aligned} \tag{5.13}$$

For a given CAPEX, the Green energy utilization maximization problem from (5.10) is,

$$\max \sum_{t=1}^{T}\sum_{b=1}^{\mathcal{B}} C^{sell}\mathcal{E}_b^{sell}(t) - \sum_{t=1}^{T}\sum_{b=1}^{\mathcal{B}} C^{sh}\mathcal{S}_b(t) - \sum_{t=1}^{T}\sum_{b=1}^{\mathcal{B}} C^{buy}\mathcal{G}_b(t). \tag{5.14}$$

Using Eqs. (5.11), (5.12), and (5.13), (5.10) can be simplified as,

$$\begin{aligned} = \sum_{t=1}^{T}\Bigg(&\sum_{b=1}^{\mathcal{B}} C^{ser}U_b(t) + (C^{buy} - C^{sell} - C^{sh})\sum_{b=1}^{ED}\mathcal{S}_b(t) \\ &+ \underbrace{C^{sell}\sum_{b'=1}^{ES}\mathcal{E}^S(b',t) - C_{buy}\sum_{b=1}^{ED}\mathcal{D}(b,t)}_{=(C^{sell}\mathcal{E}^S(t) - C_{buy}\mathcal{D}(t)),\text{ constant at } t}\Bigg). \end{aligned} \tag{5.15}$$

From (5.15), it is observed that $C_{buy}, C_{share},$ and R_{sell} depend on the CAPEX provisioning on a BS. These three parameters are influenced by changing CAPEX and depend on the quantum of energy harvest. On the contrary, C^{ser} is representative of the user QoS satisfaction in the network and doesn't depend on the CAPEX provisioning. Hence, the service operator profit maximization is broken in two parts wherein, it is first solved for user service maximization and then the problem is solved for the part containing cost parameters which are influenced by CAPEX.

From (5.15), it is inferred that the profit maximization problem contains two decision variables, U_b i.e., users being served in the network and $\mathcal{S}_b$, representing the amount of energy shared to a deficit BS in the network. Hence, maximizing user service and green energy utilization in the network leads to increasing the service operator profit. Both the decision variables are independent to each other. Hence, maximizing both these variables individually will be equivalent to maximizing them together. It may be noted that the current framework does not impose any penalty on the operator for not meeting user QoS requirements of all active users in the network. If the operator does not meet user QoS, operators loses out on service revenue. Hence, to maximize profit, operator strives to meet QoS requirements of all the network users.

5.3.1 User Service Maximization

It is assumed that the BSs are dynamically allocated the frequency spectrum depending on the fraction of users associated with them. Hence, BS b is allocated a bandwidth $BW_b = BW \times L_b(t)$, with BW denoting the total bandwidth availability. Further, each user associated with BS b is allocated equal bandwidth resource $BW_{ub} = BW_b/U_b$. The rate achievable by a user u when associated with BS b is given as,

$$\begin{aligned} r_{ub}(t) &= BW_{ub} \log_2(1 + SNR_{ub}(t)), \\ \text{with, } SNR_{ub}(t) &= \frac{\mathcal{P}_{ub}(t) g_{ub}(t)}{(BW_{ub}\sigma^2) d_{ub}^n} \geq \theta_{th}. \end{aligned} \tag{5.16}$$

Here, $\mathcal{P}_{ub}$ denotes the downlink power transmitted by BS b to user u, g_{ub} denotes the channel gain between user u and BS b, σ^2 denotes PSD of additive white Gaussian noise, d_{ub} denotes the physical distance between the user and BS, n denotes the path loss exponent, and θ_{th} signifying the QoS requirement per user.

It may be noted that, in this chapter, the user QoS is defined as the minimum signal to noise ratio, SNR, required by a user when associated with a BS, θ_{th}. Hence, a user is considered to be served by the network, only if the user achieves SNR higher than θ_{th} when associated with a BS.

The user service maximization problem is given as,

$$\begin{aligned} P_1 : \max_{U_b} \sum_{t=1}^{T} \sum_{b=1}^{\mathcal{B}} C^{ser} U_b(t) \\ s.t., \; U_b : SNR_{ub} \geq \theta_{th}, u \in \{U_b\} \forall b \in \mathcal{B}. \end{aligned} \tag{5.17}$$

The problem P_1 is equivalent to maximizing the instantaneous sum rate of the network given as,

$$
\begin{aligned}
P_2 : \max_{\pi_{ub}, \mathcal{P}_{ub}} & \sum_{b=1}^{\mathcal{B}} \sum_{u=1}^{U} \pi_{ub} BW_{ub} \log_2 \left(1 + \frac{\mathcal{P}_{ub} g_{ub}}{(BW_{ub} \sigma^2 d_{ub}^n)}\right) \\
s.t., & C1. \pi_{ub} \in \{0, 1\} \xrightarrow{\text{relaxed as}} \pi_{ub} \in [0, 1] \\
& C2. \sum_{b=1}^{\mathcal{B}} \pi_{ub} = 1, \\
& C3. \pi_{ub} \geq 0, \\
& C4. \mathcal{P}_b = \sum_{u} \mathcal{P}_{ub} \leq \mathcal{P}_{max} \\
& C5. \mathcal{P}_{ub} \geq \theta_{th} BW_{ub} \sigma^2 d_{ub}^n / g_{ub}.
\end{aligned} \tag{5.18}
$$

Constraint $C1$ indicates that a user can either be associated with a BS or be out of service. $C2$ indicates that at a time a user can be associated with only one BS. $C4$ constraints the power allocation by a BS to a upper limit $\mathcal{P}_{max}$ as per the FCC limits. $C5$ constraints the network to meet the user QoS. Due to constraint $C1$, this problem is a 0-1 knapsack combinatorial optimization problem, and is NP hard in nature. To solve the problem, C1 is relaxed such that, $\pi_{ub} \in \{0, 1\} \longrightarrow \pi_{ub} \in [0, 1]$, i.e., the space of the optimization variable is changed from integer to real, $\pi_{ub} \in \mathbb{Z}^+ \longrightarrow \mathbb{R}^+$. After performing this relaxation, the objective function of problem P_2 is now a convex problem with affine constraints. It may be noted that $C5$ denotes the user QoS constraint given in (5.16). The Lagrangian to P_2 is,

$$
\begin{aligned}
\mathbb{L}(\underline{\pi}, \underline{P}, \underline{\alpha}, \underline{\beta}, \underline{\gamma}, \underline{\delta}) = & \sum_{b=1}^{\mathcal{B}} \sum_{u=1}^{U} \pi_{ub} BW_{ub} \log_2 \left(1 + \frac{\mathcal{P}_{ub} g_{ub}}{(BW_{ub} \sigma^2 d^n ub)}\right) \\
& + \sum_{u} \alpha_u \left(\sum_{b} \pi_{ub} - 1\right) \\
& + \sum_{b} \gamma_{ub} \left(\sum_{u} \mathcal{P}_{ub} - \mathcal{P}_{max}\right) - \sum_{b} \sum_{u} \beta_{ub} \pi_{ub} \\
& - \sum_{u} \delta_u (\mathcal{P}_{ub} - \theta_{th} BW_{ub} \sigma^2 d_{ub}^n / g_{ub}).
\end{aligned} \tag{5.19}
$$

Solving the Lagrangian, the below expressions are obtained,

$$
\begin{aligned}
\sum_{u} \mathcal{P}_{ub} = \mathcal{P}_b = & \frac{\exp \sum_u \left(\frac{(\beta_{ub} - \alpha_u) \ln 2}{BW_{ub}}\right) - U_b}{\sum_u g_{ub} / (BW_{ub} \sigma^2 d^n ub)} \\
\text{or, } \mathcal{P}_b^* = \min & \left\{ \frac{\exp \sum_u \left(\frac{(\beta_{ub} - \alpha_u) \ln 2}{BW_{ub}}\right) - U_b}{\sum_u g_{ub} / (BW_{ub} \sigma^2 d^n ub)}, \mathcal{P}_{max} \right\},
\end{aligned} \tag{5.20}
$$

$$\text{and,} \left(\sum_u \pi_{ub}\right)^* = U_b^* = \sum_u \left(\frac{(\delta_u - \gamma_{ub}) \ln 2 g_{ub}}{BW_{ub}^2 \sigma^2 d_{ub}^n}\right). \tag{5.21}$$

It can be observed that the optimal values of the objective variables are a function of the Lagrange-dual multipliers. In order to compute these dual variables, sub-gradient based search [128] is used to compute and update the variables, as

$$\alpha_u(i+1) = \alpha_u(t) - S_1[\sum_b \pi_{ub} - 1] \tag{5.22}$$

$$\beta_{ub}(i+1) = \beta_{ub}(t) - S_2[\pi_{ub}] \tag{5.23}$$

$$\gamma_b(i+1) = \gamma_b(t) - S_3[\mathcal{P}_b - \mathcal{P}_{max}] \tag{5.24}$$

$$\delta_u(i+1) = \delta_u(t) - S_4[BW_{ub}\log_2(1 + \mathcal{P}_b\theta_1(u)) - r_0]. \tag{5.25}$$

Here, S_j represents the step size used to update the dual variables, and i is the number of iterations. Problem P_2 was based on instantaneous sum rate maximization. To study the effects of averaging the stochastic fading channel, an expected rate maximization problem is formulated. From (5.16), the only parameter introducing randomness is the channel gain g_{ub}. The channel between the user u and the BS b is assumed Rayleigh fading, resulting in the power gain to be exponentially distributed. Let $\mathcal{P}_{ub}/(BW_{ub}\sigma^2 d_{ub}^2) = K$.

Lemma 5.1 *PDF of rate achievable by a user over a Rayleigh fading channel is*

$$f_R(r) = \exp\left\{1 - \exp\left(\frac{r \ln 2/BW_{ub})}{K}\right)\right\}\left(\frac{\ln 2 e^{r \ln 2/BW_{ub}}}{BW_{ub} \times K}\right). \tag{5.26}$$

Proof

$$\begin{aligned} F_R(r) &= \mathbb{P}[R \le r] = \mathbb{P}(r_{ub} \le r) \\ &= \mathbb{P}\left(g_{ub} \le \frac{e^{r \ln 2/BW_{ub}} - 1}{K}\right) \end{aligned} \tag{5.27}$$

Using transformation of random variables, the PDF of rate achievable is obtained as shown in (5.26). □

Using the derived PDF, the expected rate comes to be

$$\mathbb{E}_R(r) = \frac{BW_{ub}}{\ln 2} e^{1/K} E_i(1/K) \tag{5.28}$$

Here, $E_i(.)$ represents the euler integral. This expression can be approximated [28] as

$$\mathbb{E}_R(r) = \frac{BW_{ub}e^{-1/K}}{\ln 2}\left(-E + \ln(1/K) + 1/K\right), \tag{5.29}$$

with E being Euler's constant ($E = 0.5772157$). The modified expected rate maximization formulation is shown below.

$$\begin{aligned} P_3 :\ & \max_{\pi_{ub}, \mathcal{P}_{ub}} \sum_{b=1}^{\mathcal{B}} \sum_{u=1}^{U} \pi_{ub} \frac{BW_{ub}e^{-1/K}}{\ln 2}\left(-E + \ln(1/K) + 1/K\right) \\ & s.t.,\ C1 - C4, \\ & C5.\ \mathcal{P}_{ub} \geq \theta_{th} BW_{ub} \sigma^2 d_{ub}^2. \end{aligned} \tag{5.30}$$

It may be noted that $C5$ now denotes the expected signal to noise ratio (SNR) constraint. The sub-gradient based search discussed before is again used to solve P_3. The green energy maximization problem is discussed in the next subsection.

5.3.2 Green Energy Utilization Maximization

The revenue maximization problem at a given CAPEX can be formulated as,

$$\begin{aligned} \mathcal{P}3 :\ & \max_{\mathcal{E}_{b'}^{sell}, \mathcal{S}_b, \mathcal{G}_b} \left(C^{sell} \sum_{b'=1}^{ES} \mathcal{E}_{b'}^{sell}(t) - C^{sh} \sum_{b=1}^{ED} \mathcal{S}_b(t) - C^{buy} \sum_{b=1}^{ED} \mathcal{G}_b(t) \right) \\ \text{s.t. } C1 :\ & \sum_{b=1}^{ED} \mathcal{S}_b(t) \leq \sum_{b=1}^{ED} \mathcal{D}(b,t) \\ C2 :\ & \sum_{b=1}^{ED} \mathcal{S}_b(t) < \min \left\{ \sum_{b'=1}^{ES} \mathcal{E}^S(b',t), \sum_{b=1}^{ED} \mathcal{D}(b,t) \right\} \\ C3 :\ & \sum_{b=1}^{ED} \mathcal{S}_b(t) \geq 0 \\ C4 :\ & \sum_{b=1}^{ED} \mathcal{G}_b(t) = \sum_{b=1}^{ED} \mathcal{D}(b,t) - \sum_{b=1}^{ED} \mathcal{S}_b(t) \\ C5 :\ & \sum_{b'=1}^{ES} \mathcal{E}_{b'}^{sell}(t) = \sum_{b'=1}^{ES} \mathcal{E}^S(b',t) - \sum_{b=1}^{ED} \mathcal{S}_b(t) \\ C6 :\ & \sum_{b'=1}^{ES} C^{sell} \mathcal{E}_{b'}^{sell}(t) \geq 0. \end{aligned} \tag{5.31}$$

It is observed that the problem $\mathcal{P}3$ has three decision variables, $\mathcal{E}_{b'}^{sell}, \mathcal{S}_b, \mathcal{G}_b$. Also, it is observed that constraint $C1 \subset C2$, thus eliminating $C1$.

For $\sum_{b=1}^{ES} \mathcal{E}_b^S(t) > \sum_{b=1}^{ED} \mathcal{D}(b,t)$, the amount of energy which can be sold back to the grid can be expressed in terms of $\mathcal{S}_b(t)$ through $C5$. Similarly, the network grid energy procurement $\sum_{b=1}^{ED} \mathcal{G}_b(t)$ can be expressed in terms of $\sum_{b=1}^{ED} \mathcal{S}_b(t)$ through $C4$. Therefore, a three variable problem $\mathcal{P}3$ is converted into a single variable problem $\mathcal{P}4$ as

$$\begin{aligned}
\mathcal{P}4 : \max_{\mathcal{S}_b} & \left(C^{sell} \left(\sum_{b=1}^{ES} \mathcal{E}_b^S - \sum_{b=1}^{ED} \mathcal{S}_b(t) \right) - C^{sh} \sum_{b=1}^{ED} \mathcal{S}_b(t) \right. \\
& \left. -C^{buy} \left(\sum_{b=1}^{ED} \mathcal{D}(b,t) - \sum_{b=1}^{ED} \mathcal{S}_b(t) \right) \right) \\
= \max_{\mathcal{S}_b} & \left((C^{buy} - C^{sell} - C^{sh}) \sum_{b=1}^{ED} \mathcal{S}_b(t) + C^{sell} \sum_{b=1}^{ES} \mathcal{E}_b^S - C^{buy} \sum_{b=1}^{ED} \mathcal{D}(b,t) \right).
\end{aligned} \tag{5.32}$$

Again, in order to derive closed form expressions and to make the problem mathematically tractable, the problem formulated in (5.32) is revised by squaring the terms. Also, since this formulation is at a given hour t, the variable t is removed from the formulation for visual clarity. The revised convex problem formulation $\mathcal{P}5$ is shown below.

$$\begin{aligned}
\mathcal{P}5 : \max_{\mathcal{S}_b} & \left((C^{buy} - C^{sell} - C^{sh})^2 \sum_{b=1}^{ED} (\mathcal{S}_b)^2 + \underbrace{(C^{sell})^2 \sum_{b=1}^{ES} \left(\mathcal{E}_b^S\right)^2 - (C^{buy})^2}_{\text{constant, } \kappa} \right) \\
\text{s.t. } C1 : & \sum_{b=1}^{ED} \mathcal{S}_b < \min \left\{ \sum_{b=1}^{ES} \mathcal{E}_b^S, \sum_{b=1}^{ED} \mathcal{D}_b \right\} \\
C2 : & \sum_{b=1}^{ED} \mathcal{S}_b \geq 0.
\end{aligned} \tag{5.33}$$

$\mathcal{P}5$ is a convex problem as the objective function is convex, constraint C1 is concave, and constraint C2 is affine. The second and third terms of the above objective function are constant and independent of the decision variable $\mathcal{S}_b$ in an hour t and hence are expressed as a constant κ.

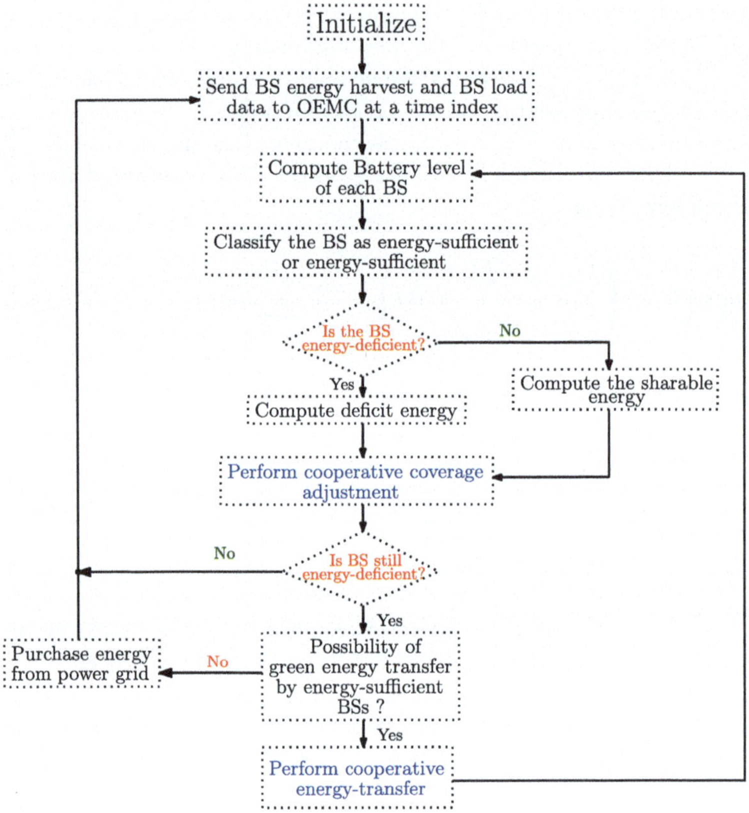

Fig. 5.2 Illustration of the JTEB framework

Theorem 5.1 *For a given CAPEX, the OPEX incurred in the grid energy procurement minimization problem is identical to the operator revenue maximization problem. That is, the solution of* $\mathcal{P}5$ *is*

$$\mathcal{S}_b(t) = \begin{cases} \mathcal{D}(b,t), \ \textit{if} \ \sum_{b=1}^{ED} \mathcal{D}(b,t) \leq \sum_{b'=1}^{ES} \mathcal{E}^S(b',t) \\ \left(\mathcal{D}(b,t) \sum_{b'=1}^{ES} \mathcal{E}^S(b',t)\right) / \sum_{b=1}^{ED} \mathcal{D}(b,t), \\ \qquad \textit{if} \ \sum_{b'=1}^{ES} \mathcal{E}^S(b',t) < \sum_{b=1}^{ED} \mathcal{D}(b,t). \end{cases} \tag{5.34}$$

Proof Please refer to Appendix C. □

The working of the joint traffic-energy balancing (JTEB) framework is briefly described below, as illustrated through Fig. 5.2. The JTEB framework involves the BSs first sending the energy harvest data and BS load data to the OEMC through optic fiber based backhaul lines, following which the battery level of each BS is computed. The OEMC then classifies each BS as energy-deficient if the battery level is below an operator defined threshold or energy-sufficient if the battery level of the BS is above the operator defined threshold.

Depending on the BS class, the deficit energy required per energy-deficient BS or excess sharable energy with each energy-sufficient BS after serving its current load is computed. The JTEB strategy first performs cooperative coverage adjustment among the networked BSs to alleviate the load of energy-deficient BSs. In the cooperative coverage adjustment process, the OEMC checks if the neighboring BSs of the energy-deficient BS have capacity to offload some users. If the neighboring BSs can offload and cater to some users of the energy-deficient BS, then the user association is changed and cellular coverage areas are modified accordingly.

After performing cooperative coverage adjustment, there might still be a possibility that due to the reconfigured BS loads, some BSs might have some additional green energy in there respective energy storage. Hence cooperative energy sharing is performed by the OEMC to ensure complete utilization of green energy at that time instant. If no BS requires aid from the energy-sufficient BSs then the remaining sharable energy is sold to the power grid. In the event where there is no sharable energy remaining with any energy-sufficient BS, but the energy-deficient BSs still require energy to meet user QoS, then energy is procured from the power grid. This cycle is repeated at each time instant, so as to minimize the grid energy procurement and guarantee user QoS to the full extent. The complexity of the JTEB framework is in the order of $\mathbb{O}(T \times \mathcal{B}^2)$, i.e., the complexity of the framework is linear with time, and increases quadratically with the number of BSs.

It may be noted that the amount of users being served by a BS is limited by the FCC regulation on antenna power radiation. Similarly, the amount of green energy transferred in the network is limited by the storage battery capacity with each BS. Thus, the operator profit for a given CAPEX provisioning can not be infinite and is limited by these two physical practical limitations.

5.4 Results

Through this section, the key results and observations are presented. As state-of-art, the performance of the JTEB framework is compared with the baseline without JTEB (without JTEB), only coverage adjustment based traffic balancing (only-TLB) and, only energy balancing (only GEB) based frameworks. The frameworks are compared on the basis of user QoS satisfaction (Fig. 5.3), sharable green energy (Fig. 5.4a), CAPEX provisioning to achieve sustainability (Fig. 5.4b), and net service operator profit (Fig. 5.5).

The simulations have been performed in MATLAB R-2022b and the simulation parameter values are provided in Table 5.1.

The simulations have been performed using annual solar harvest data of New Delhi city, obtained from National Renewable Energy Laboratory [112]. The BSs are assumed to be in the same city, hence for result generation and simulations, the energy harvest at each BS has been assumed to be equal. The network dynamics

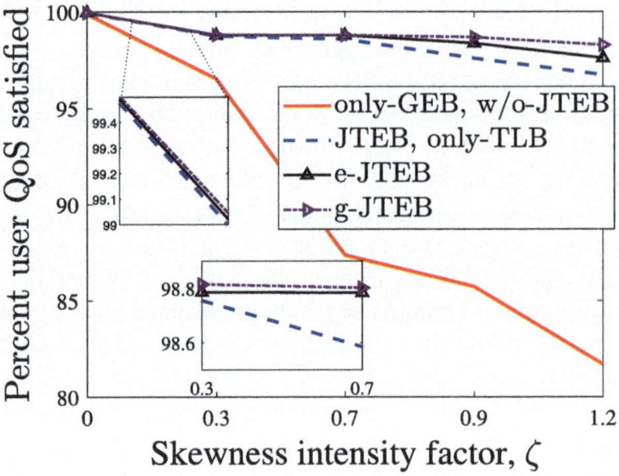

Fig. 5.3 Illustrating variation of user QoS satisfaction with skewness intensity factor in the JTEB, expected-JTEB (e-JTEB), global-JTEB (g-JTEB), and other competitive state of art frameworks

arise due to BS load variations, which result in variation of BS power consumption and BS green energy storage levels. The skewness intensity factor introduced in Chapter IV captures the extent of network dynamics.

It may be highlighted while the analytical framework is generalized, for simulation studies, the cellular traffic data [114] and the energy harvest data [112], available in literature are hourly in nature. Hence, the BS operations namely, coverage adjustment, energy transfer, and energy trade with the grid are performed on an hourly basis by the radio network controller, resulting in update of the BS parameters on an hourly basis.

The skewness intensity factor ζ modeled through (4.1) represents the fraction of higher traffic a BS experiences as compared to the balanced load scenario (when $\zeta = 0$). As ζ increases, any random BS in the network experiences a higher skewed fraction of traffic. The system has been simulated up to $\zeta = 1.2$, which corresponds to around 80% higher traffic than the balanced scenario. While ζ can take more higher values, the current study has been restricted to 80% traffic imbalance. Skewness levels higher than $\zeta = 1.2$ are not practical and have a low probability of occurrence in reality.

5.4.1 User QoS Performance

User QoS satisfaction is measured on the basis of number users whose QoS is being guaranteed and is depicted in Fig. 5.3. The problems formulated in P_2 (JTEB) and P_3 (expected JTEB, e-JTEB) were relaxed to make the problems tractable. Exhaustive search has been used to compute the global optimum user QoS performance satisfaction (g-JTEB). Qualitatively the nature of the plots in Fig. 5.3 for the JTEB, e-JTEB, and g-JTEB are similar to each other. Quantitatively, it is

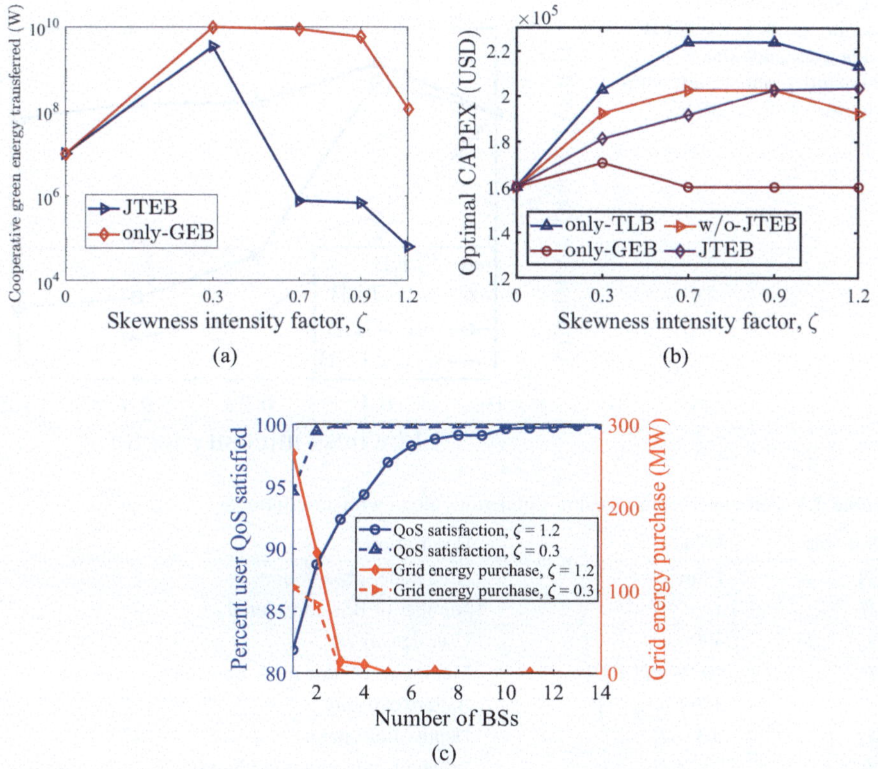

Fig. 5.4 (**a**) Green energy utilization as a function of traffic-energy skewness; (**b**) optimal CAPEX for energy sustainability versus traffic-energy skewness; (**c**) user satisfaction and grid energy requirement as a function of networked BS cluster size

observed that the expected sum rate maximization results in a marginally better performance than the instantaneous sum rate solution especially at higher skewness conditions. This behavior may be attributed to the fact that since the framework analyzes from a BS perspective, expected sum rate maximization results averaging out the stochastic channel behavior, resulting in marginally higher performance especially at higher skewness. Both the instantaneous and expected rate solutions perform marginally poor than the g-JTEB, which is essentially due to the constraint $C1$ relaxation. It is also inferred from Fig. 5.3 that coverage adjustment based methods perform significantly better than non-coverage adjustment frameworks (only-GEB, w/o-JTEB), and the performance is more profound with increasing skewness factor (up to 18.38% gain at extreme skewness, $\zeta = 1.2$).

Hence, it is inferred that the coverage adjustment based JTEB framework provides significant improvement in network user QoS performance over other competitive approaches.

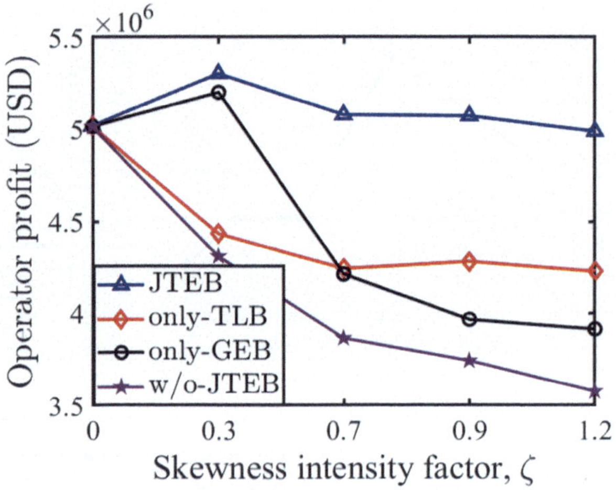

Fig. 5.5 Variation of operator profit in the JTEB framework and other competitive state of art, with a cluster of seven BSs

Table 5.1 Parameter values used in simulations along with description

Variable	Value	Description
$\mathcal{A}$	1 km^2	Area under observation
$\mathcal{B}$	7	Number of BSs considered
λ_u	3000	User density
$\mathcal{P}_{max}$	40 W [80, 109]	Maximum BS downlink transmit power level
$\mathbb{B}_c$	2460 Wh [67]	Battery capacity
δ	0.3 [67]	Depth of discharge
BW	20 MHz [129]	System bandwidth considered
σ^2	−150 dBm/Hz [130]	PSD of AWGN
r_0	300×10^3 bps [129]	User QoS
p_0	0.9 [131]	Non-outage probability
C_{PV}	1300 USD [132]	Cost of unit PV panel
C_B	216 USD [133]	Cost of unit storage battery
C^{sh}	0.015 USD	Price of sharing unit energy
C^{buy}	0.079 USD [134]	Price of purchasing unit energy from power grid
C^{sell}	0.057 USD [135]	Price of selling unit energy

5.4.2 *Sustainable Design*

Through Fig. 5.4a the cooperative green energy shared in the network among the BSs is illustrated, thereby improving the temporal green energy utilization through the JTEB framework. It is inferred that amount of green energy shared in the JTEB framework decreases with increasing relative skewness of the BS loads. it is also inferred that the JTEB framework performs poorly as compared to the 'only-GEB' framework. In this regard, it may be noted that Fig. 5.4a should be analyzed together with Fig. 5.3, wherein it is observed that the user QoS performance of the JTEB framework is much better than the 'only-GEB' framework. Thus, signifying that the loss in green energy shared is actually being used up in improving the user QoS.

Figure 5.4b shows the variation of optimal solar CAPEX required to be provisioned to achieve energy sustainability with increasing relative skewness of load inhomogeneity. It is observed that the 'only-GEB' framework attains sustainability with the least CAPEX. This is because the 'only-GEB' involves a significant amount of green energy sharing (as depicted in Fig. 5.4a), which results in reducing the CAPEX required to achieve sustainability. It may be noted that while the only-GEB framework incurs least CAPEX, the user QoS is severely affected with increasing traffic inhomogeneity (Fig. 5.3).

It is observed that at lower skewness the JTEB framework incurs much lesser CAPEX to achieve energy sustainable network over 'without JTEB' framework, while as the relative skewness increases the optimal CAPEX required for the JTEB is marginally higher than the 'without JTEB' strategy. This behavior in the nature of optimal CAPEX plot is explained as follows. At lower skewness, the JTEB incurs lesser CAPEX than the baseline, i.e., 'without JTEB' because cooperative energy sharing results in improving the green energy utilization within the network, thus reducing the CAPEX. As the relative skewness increases, it is observed from Fig. 5.3 that the JTEB strategy results in serving a higher number of users in the network, thus improving the user QoS satisfaction. Due to catering to higher users in the network, the JTEB framework experiences lesser green energy cooperative sharing (as illustrated through Fig. 5.4a). On the contrary, the 'without JTEB' framework is observed to result in much lesser user QoS guarantee in the network, resulting in lower CAPEX as compared to JTEB at higher relative skewness levels to attain energy-sustainability.

Figure 5.4c depicts the variation of user satisfaction and grid energy procurement with number of BSs at skewness levels, $\zeta = 0.3, 1.2$. While the user QoS satisfaction curves have an increasing nature, the grid energy procurement curve is observed to follow a decreasing nature, with increasing number of BSs in the network. The cooperative JTEB framework results in formation of self sustainable BS clusters. From Fig. 5.4c it is inferred that at higher skewness level, more BSs are required to cooperate with each other in terms of network services like load and energy management. For instance, when the network is subjected to a higher load inhomogeneity of $\zeta = 1.2$, around 13 BSs are required to meet 100% network user QoS alongside achieving net zero grid energy procurement. On the contrary, at $\zeta = 0.3$, around 8 BSs are sufficient to meet the user QoS in addition to forming a sustainable BS cluster.

Hence, it is inferred that the JTEB framework offers appreciable CAPEX reduction while achieving a green energy sustainable system in addition to significant user QoS gains.

5.4.3 Operator Profitability

Finally, Fig. 5.5 illustrates the variation of operator net profit with increasing traffic inhomogeneity. It is inferred that the JTEB framework significantly outperforms the baseline w/o-JTEB framework by providing gain up to 39.38% at extreme skewness level. It is observed that the JTEB initially competes with the only-GEB framework at lower skewness levels, but at higher skewness levels JTEB competes with the only-TLB framework. One reason may be attributed to the fact that at lower skewness levels, in the JTEB and only-GEB frameworks, there is much greater green energy transferred in the network (Fig. 5.4a), which assists in cutting down the grid energy procurement thereby reducing the OPEX as well as CAPEX (Fig. 5.4b). As the skewness levels increase, the only-GEB framework (in the absence of load management) loses out on user service revenue (Fig. 5.3). At higher skewness levels, the only-TLB framework loses out to the JTEB framework due to higher grid purchase in the absence of energy transfer. The JTEB framework is observed to provide gains up to 13.49%, 24.21%, and 39.38% respectively, over the only-TLB, only-GEB, and w/o-JTEB frameworks at extreme skewness levels.

Hence, it is inferred that as compared to the balanced network scenario, at higher skewness levels, the joint traffic and energy optimization based JTEB framework provides superior profits. Whereas in other state-of-art frameworks, it is observed that the operator profit degrades significantly as compared to the balanced load scenario.

5.5 Summary

This chapter has showcased design of operator revenue aware telecommunication networks, through a joint load and energy balancing framework. The system performance evaluation has been computed through user QoS and operator profit. The revenue metrics associated with designing communication network have been analyzed. A joint load and energy balancing framework has been shown to maximize operator revenue by maximizing user service and green energy utilization. Our analysis has shown that energy balancing assists in complementing load-balancing, and is beneficial in scaling green networking solutions.

Chapter 6
Exploring Aerial-Terrestrial Future Networks

6.1 Introduction

Non-terrestrial networks are a key theme in the upcoming 6G communication technology [136]. Recent research have pertained around integrating the existing terrestrial communication networks involving cellular communication frameworks with aerial and space networks [101, 137]. Aerial technologies for instance, HAPS and UAVs are widely being used as a 3D BS or aerial BS, thus bring mobility and flexibility to the conventional terrestrial communication framework [138, 139]. Outer space networks through deployment of LEO satellites, are also expected to be used in urban areas, to improve the communication network connectivity and coverage. Drone based UAVs are deployed up to heights of around 1 km from the ground level, while satellites are deployed in the orbit at heights of at least beyond 400 km. Further, drone based UAVs are energy constrained devices and thus have a limited flight time.

Considering the challenges in drone based UAVs and satellites, HAPS are emerging as a very attractive technology to integrate with the terrestrial networks as it is averse to energy constraints and latency challenges occurring in drone based UAVs and satellites, respectively [105, 106]. HAPS are deployed at heights of around 20 km in the stratosphere. At the stratospheric height, the HAPS receive abundant of sunlight being above the clouds, in addition to low wind speed, thus aiding in a stable deployment of the HAPS. Additionally, the HAPS mounted BS, hMBS, can operate in the sub-6 GHz, i.e., LTE bands, as well as mmWave bands, thus having the potential to integrate with the existing communication framework.

As a case study for energy-efficient non-terrestrial networks, this chapter explores designing integrated HAPS aided terrestrial communication framework, from the aspect of profitability to the mobile service provider. The communication and energy networks in the considered system are studied and modeled jointly as a DTMC. In the event of low energy harvest or high traffic density being

L.-C. Wang et al., *Green Energy Driven Integrated Smart Grid and Wireless Networks*, Wireless Networks, https://doi.org/10.1007/978-3-032-20509-4_6

experienced by the tMBS, with the aim of realizing a green network, the proposed framework aims to offload excess users with the tMBS to the hMBS, rather than procuring energy from the grid. The research problem is motivated in the upcoming subsection.

6.1.1 Motivation

Solar powered and smart grid connected, "dual-powered" tMBSs are becoming very attractive as a cost and energy efficient network solution to mobile operators. Despite being dual-powered, they are not carbon free. It is very challenging to gauge the cellular traffic density in a tMBS cell, as the cellular traffic cannot be controlled. Additionally, solar powered tMBSs are prone to intermittent energy harvest, resulting in sudden energy outages if not connected with power grid. Hence, the user QoS gets affected due to randomness in cellular traffic density and energy harvest.

In addition to integrating aerial and terrestrial networks, achieving energy sustainable terrestrial network design has been a key objective of green communication [2]. With the terrestrial macro BSs, tMBSs, being traditionally powered with carbon emitting power grids, there has been research around powering the tMBSs with purely renewable energy sources (like solar energy) [140]. While such solutions are green, they are not cost effective as well as scalable solutions to the mobile operator due to the high capital investment involved.

It may also be noted that the tMBSs are governed by FCC guidelines on downlink power radiation [80]. Thus, a tMBS can meet the QoS guarantee of limited cellular users, in the event of high traffic density or low energy harvest. Hence, there is a requirement for additional capacity injection, especially in an urban/dense urban scenario. Heterogeneous networks involving small cells have been a traditional strategy, but they result in extra deployment cost to the operator and are not green. HAPS mounted radio access networks which can provide coverage up to 500 km, are being studied as an attractive solution. As hMBS provides a wide coverage, the capacity of hMBS per cell is limited [105]. Thus, there is requirement to study the improvement of network QoS under energy and capacity limited hMBS offloading capabilities. In a hMBS aided, solar powered, and smart grid connected communication network, by characterizing the communication and energy networks as a DTMC, this chapter aims to study user offloading as an alternative to grid energy procurement in the event of higher user density or low energy harvest, towards realizing operator cost profitable green networks. In the upcoming subsections the key chapter contributions and organization is discussed.

6.1.2 Key Insights from This Chapter

The key insights from this chapter are as follows:

1. We showcase a HAPS aided, solar powered, and smart grid connected green communication network. From a system perspective, to realize green networks, the framework aims to offload excess users to the hMBS in the event of low energy harvest or high user density rather than relying on grid energy procurement. The power grid infrastructure is utilized only for selling surplus green energy harvested back to the power grid.
2. The inherent communication and energy network in the proposed system are studied and jointly modeled as a six state DTMC. Through the proposed framework, the work explores the extent of user offloading to an energy and capacity constrained hMBS.
3. We also analyze the solar provisioning required at the hMBS for radio access network (RAN) functioning, and the operator deployability aspects which are crucial to determine the sustenance of non-terrestrial networks.

6.1.3 Organization

The chapter layout is as follows: Sect. 6.2 outlines the system model in detail. Section 6.3 presents the proposed joint communication-energy network Markovian model. Section 6.4 presents the operator profit analysis and sustainable system design, while Sect. 6.5 discusses the results and observations. Section 6.6 concludes the chapter.

6.2 System Model

The work considers a downlink two-tier wireless RAN consisting of solar powered and smart grid connected tMBS and an aerial hMBS as shown in Fig. 6.1a. The tMBS is solar provisioned to harvest green energy with PV panels and storage batteries. The tMBS uses the grid connectivity to sell temporally surplus green energy available with it and is assumed to not procure any energy from power grid. A closed area $\mathcal{A}$, consisting of a set of $\mathcal{U}$ active users denoted as $\mathbf{U} \in \{1, \ldots, \mathcal{U}\}$ modeled as a homogeneous binomial point process, is considered. The user set varies temporally as discussed in the later subsections. The users can be served either by the tMBS or the hMBS represented as, $\mathbf{B} \in \{B_1, B_2\}$. It may be noted that while the current work considers a single tMBS framework, the analysis can be generalized to any number of tMBSs. It may also be noted that B_2 represents a hMBS.

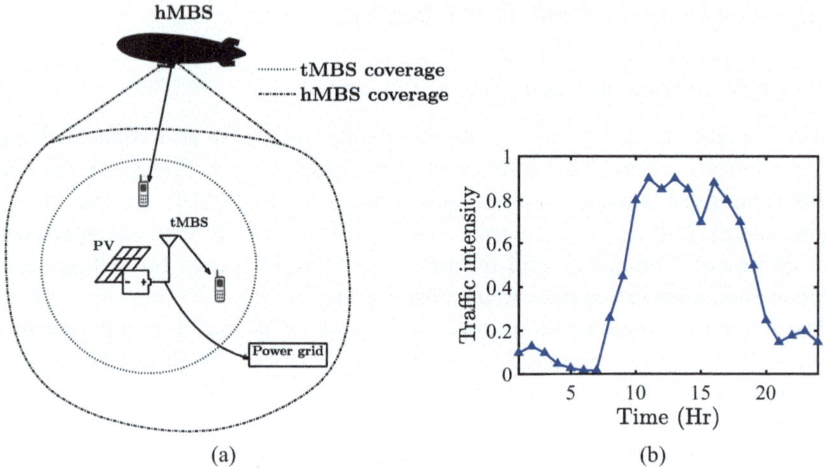

Fig. 6.1 (**a**) Illustration of HAPS aided, grid connected, and solar powered communication network, (**b**) Time varying traffic profile in the considered area [2]

The hMBS is assumed to be self-sustainable, being equipped with sufficient solar panels and storage batteries required to sustain the HAPS in stratosphere [141]. The current chapter assumes that the hMBS operates in the sub-6 GHz frequency spectrum with a single beam, and is equipped with an uniform planar array (UPA) antenna structure to mitigate the increased free space path loss. With the total bandwidth resource being W, the tMBS is assumed to operate in full frequency reuse mode, having access to the entire frequency resource. The hMBS is assumed to have limited capacity [105], having access to a fraction $f\%$ of this net frequency resource for the area under consideration. In the upcoming subsections, the traffic profile, energy profile, and the channel models between the hMBS/tMBS and users are discussed.

6.2.1 Traffic and Energy Harvest Profile

The tMBS is assumed to be subjected to time varying traffic $\rho(t)$ as shown in Fig. 6.1b. A tMBS $b \in \mathbf{B}$, harvests hourly $\mathcal{H}_b(t)$ green energy through $\mathcal{N}_{PV}$ PV panels which is stored in η_B storage batteries equipped with it. The annual hourly energy harvest at New Delhi city, India, has been obtained from National Renewable Energy Laboratory [142]. The battery storage has limited capacity $\beta^{max} = (\eta_B \times \beta^{cr})$ and a critical threshold level $\beta_{cr} = (\delta \times \eta_B \times \beta^{cr})$, with β^{cr} and δ denoting unit battery capacity and depth of discharge, respectively.

6.2.2 Channel Models

In this subsection the analytical modeling of the link between the tMBS–user and hMBS–user is discussed.

6.2.2.1 tMBS to User Link

The channel state information ψ_{ub} between user u and tMBS b is assumed to be Rayleigh distributed, with the corresponding channel gain $g_{ub} \in \{\mathbf{g}\}$ being exponentially distributed with unit mean. The rate achievable by a user through the associated tMBS is given as

$$r_{ub} = W_{ub} \log_2 (1 + SINR_{ub}),$$
$$\text{where, } SINR_{ub} = \frac{\mathcal{P}^T_{ub} g_{ub}}{I + (W_{ub}\sigma^2) d^2_{ub}}. \tag{6.1}$$

Here, $\mathcal{P}^T_{ub} \in \{\mathbf{P}\}$ and $d_{ub} \in \{\mathbf{d}\}$ correspond to the transmit power and distance between by tMBS b to user u, σ^2 denotes the power spectral density of additive white Gaussian noise, and I refers to the inter cell interference given as $I = (\sum_{b=1,b\neq b'}^{2} \mathcal{P}^T_{u'b'} g_{u'b'} d^{-2}_{u'b'})$. It is assumed that each user has a minimum rate guarantee QoS requirement given as r_{th} such that $r_{ub} \geq r_{th}$.

6.2.2.2 hMBS to User Link

The path loss experienced by a user when associated with a hMBS is modeled as per the 3GPP standards [143]. In general, the received power at user is represented as,

$$P_r[dB] = \mathcal{P}^T_{uH}[dB] + G^H_t[dB] + G_r[dB] - PL^H[dB]. \tag{6.2}$$

Here, $\mathcal{P}^T_{uH}$ denotes the transmit power of hMBS to user u associated with it, $PL^H = (p_{LoS} PL_{LoS} + p_{NLoS} PL_{NLoS})$ represents the average path loss at user comprising of the LoS and non-LoS components, p_{LoS} and p_{NLoS} denote the probability of LoS and non-LoS components respectively, depending on the elevation angle between hMBS and the user. The LoS and non-LoS path loss components are further modeled through the equations below.

$$PL_{LoS} = PL^b_{LoS} + PL_{ag} + PL_s + PL_e, \tag{6.3}$$

$$PL_{NLoS} = PL^b_{NLoS} + PL_{ag} + PL_s + PL_e, \tag{6.4}$$

$$\text{where, } PL^b_{LoS} = FSPL + CL_{LoS} + X_{LoS}, \tag{6.5}$$

$$PL^b_{NLoS} = FSPL + CL_{NLoS} + X_{NLoS}, \tag{6.6}$$

$$PL_s = PF/\sqrt{2} = \frac{1.1(f_c/4)^{1.5}}{\sqrt{2}}, \tag{6.7}$$

$$FSPL = 32.45 + 20\log_{10}(f_c) + 20\log_{10}(d), \tag{6.8}$$

$$\text{and } d = \sqrt{R_e^2 sin^2(\theta) + H_{hap}^2 + 2H_{hap}R_e} - R_e sin(\theta). \tag{6.9}$$

Here, PL_x^b (for $x \in \{LoS, NLoS\}$) denotes the basic path loss, PL_{ag} represents the loss due to atmospheric gases. It is neglected as the system operates in sub-6 GHz. PL_s denotes the scintillation loss at the ionosphere, while PL_e is termed as building entry loss (neglected for outdoor scenario). CL_x and X_x represent the clutter loss and shadow fading loss respectively, $FSPL$ in (4), (5) stands for free space path loss, f_c denotes the carrier frequency of operation, R_e radius of earth, θ the elevation angle, and H_{hap} denotes the height of hMBS. Depending on P_r, $r_{uh} = f\log_2(1 + SINR_{uh})$, where f is the fraction of frequency resource with the hMBS for the current area under study.

6.3 Proposed Integrated Communication-Energy Network Markovian Model

A HAPS aided, grid connected, and solar powered tMBS framework consists of a communication network and an energy network. In this section, these two systems are first presented separately and then the integrated Markovian model is elaborated in detail.

6.3.1 Communication Network Modeling

The communication network is formed by the downlink communication between tMBS/ hMBS and the users. For the set of **U** temporally active users, each having a minimum rate guarantee r_{th}, the minimum power required to be allocated per user by a tMBS is calculated as given below.

$$\begin{aligned}
&\mathbb{P}\left(r_{ub}(t) \geq r_{th}\right) \geq p_o \\
&\mathbb{P}\left(g_{ub}(t) \geq \frac{\exp\left(r_{th}\ln 2/W_{ub} - 1\right)\left(d_{ub}^2(W_{ub}\sigma^2) + I\right)}{\mathcal{P}_{ub}^T(t)}\right) \geq p_o \\
&\text{or, } \exp\left(-\frac{\exp\left(r_{th}\ln 2/W_{ub} - 1\right)\left(d_{ub}^2(W_{ub}\sigma^2) + I\right)}{\mathcal{P}_{ub}^T(t)}\right) \geq p_o \\
&\mathcal{P}_{ub}^T(t) \geq \frac{\exp\left(r_{th}\ln 2/W_{ub} - 1\right)\left(d_{ub}^2(W_{ub}\sigma^2) + I\right)}{\ln(1/p_o)}.
\end{aligned} \tag{6.10}$$

The net power consumption of the tMBS for RAN, as a function of number of users, channel gain, and distance of each user, is calculated as

$$\sum_{u} \pi_{ub} \mathcal{P}^{T}_{ub}(t) = P^{T}_{b}(t) = f(\mathbf{U}, \mathbf{g}, \mathbf{d}) \leq \mathcal{P}_{max}. \tag{6.11}$$

Here, $\mathcal{P}_{max}$ refers to the maximum transmit power level of the tMBS in accordance with the FCC guidelines. $\pi_{ub} \in \{0, 1\}$ refers to a variable indicating association of user u to BS b and is defined as

$$\pi_{ub} = \begin{cases} 1, & \text{if user } u \text{ associated with } b \\ 0, & \text{else.} \end{cases} \tag{6.12}$$

A communication system can be modeled as a two state Markov model as described below.

1. *QoS satisfied (QS)* state: The system will be in QS state at time t if the QoS rate requirement of all the active users are met by the tMBS and the net power consumption of the tMBS does not violate the FCC guidelines (6.11), i.e.,

$$r_{ub}(t) \geq r_{th} \; \forall \; u \in \mathbf{U} \text{ such that, (6.11).} \tag{6.13}$$

2. *QoS violation (QV)* state: The system will be in QV state at time t if the QoS rate requirement of the active users are not being met by the tMBS. In this state, the transmit power level of the tMBS saturates at $\mathcal{P}_{max}$ with some users $\mathbf{U}' \subset \mathbf{U}$ being unserved. Mathematically,

$$r_{ub}(t) < r_{th} \; \forall \; u \in \mathbf{U}' \subset \mathbf{U} \text{ such that, (6.11).} \tag{6.14}$$

The upcoming subsection entails the energy network in the system.

6.3.2 Energy Network Modeling

The solar provisioning at the tMBS (including the PV panels and storage batteries) and the power grid together form the energy network in the system. Depending on the green energy harvest and traffic profile being experienced by a tMBS, the energy network can be modeled as a three state system explained below. First, the green energy storage level at tMBS b is defined as

$$\begin{aligned} B'_{b}(t) &= B_{b}(t-1) + \mathcal{H}_{b}(t) - \mathcal{E}_{b}(t), \\ B_{b}(t) &= \min\{\max\{B'_{b}(t), \beta_{cr}\}, \beta^{max}\}. \end{aligned} \tag{6.15}$$

Here, $\mathcal{E}_b(t) = N_{TRX}\left(\mathcal{P}_0 + \theta_1 \mathcal{P}_b^T(t)\right)$ denotes the net energy consumption by the tMBS, $\mathcal{P}_0$ denotes the static energy consumption of the tMBS, N_{TRX} denotes number of transceivers, and θ_1 a constant [2]. Depending on the battery level, the energy network can be discretely modeled as

1. *Deficit (D)* state: In this state, $B_b'(t) < \beta_{cr}$. This state signifies that the tMBS becomes energy-deficient thereby affecting user service. In this case, the network will have to offload users to the hMBS or buy some energy from the power grid.
2. *Intermediate (I)* state: In this state, $\beta^{max} \geq B_b'(t) \geq \beta_{cr}$.
3. *Surplus (S)* state: In this state, $\beta^{max} < B_b'(t)$, signifying that the storage battery cannot store excess energy, thereby necessitating energy sell to the power grid.

In the coming subsection, the joint communication-energy network Markov modeling and service provider network operations are discussed and presented.

6.3.3 Joint Markov Model

The communication and the energy network model discussed previously are not independent of each other, as both the models are a function of the tMBS energy consumption. Thus, both the models are influenced by the tMBS power consumption variation to meet the QoS requirement of the temporally active users. Hence, there is a need to analyze the communication network and the energy network jointly through an integrated Markovian model as shown in Fig. 6.2, with the states as explained below. Before detailing the integrated Markov model, it is imperative to discuss the possible service provider network operations.

1. Energy selling back to grid: The tMBS can sell excess energy in its storage back to grid.
2. Excess user offloading to hMBS: In the event that the tMBS is unable to satisfy QoS of all temporal active users, some users will be offloaded to the hMBS.

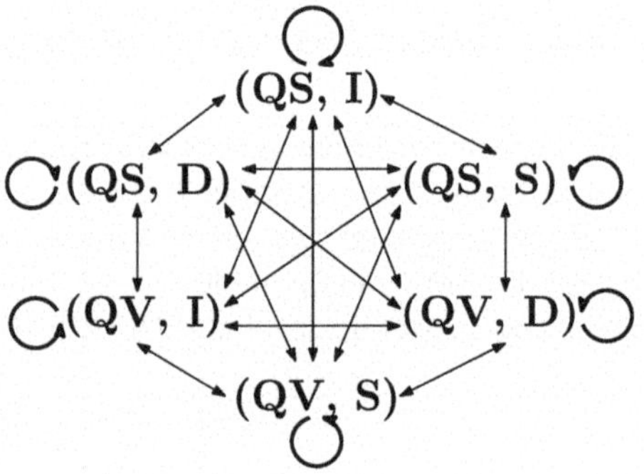

Fig. 6.2 Proposed Markovian model

It may be noted that the tMBS does not purchase energy from the grid, rather aims to offload excess users to the hMBS as required. Next, the states of the integrated Markov model are described below along with their corresponding network operations.

1. (QS, D): In this state, the tMBS satisfies QoS requirements of the temporally active users. But since battery level is at critical level, *some users require to be offloaded to the hMBS.*
2. (QS, I): The tMBS is sufficient to satisfy QoS of temporal active users and the battery level is in intermediate state, i.e., above level. *This state does not involve either of user offloading or energy selling back to grid.*
3. (QS, S): The tMBS is sufficient to satisfy QoS of temporal active users and battery level is in surplus state. *This state only involves energy selling back to the grid and does not involve user offloading to hMBS.*
4. (QV, D): The tMBS is insufficient to satisfy temporal QoS requirements, further the battery level is also in deficit state. *Hence, this state involves excess user offloading to hMBS.*
5. (QV, I): The tMBS is insufficient to satisfy temporal QoS requirements, but the battery level is not in deficit state. *Hence this state involves only excess user offloading to hMBS.*
6. (QV, S): The tMBS is insufficient to satisfy temporal QoS requirements, but battery storage is overflowing. *This state involves excess user offloading to hMBS and energy selling back to the grid.*

The transition probability matrix corresponding to this DTMC is denoted as $\mathbf{T} \in \mathbb{R}^{6\times 6}$, while the steady state probability matrix is given as $\underline{\pi} \in \mathbb{R}^{6}$. The following system of equations are solved in order to obtain the steady state probabilities,

$$\underline{\pi} = \underline{\pi}^T \mathbf{T}, \text{ such that } \sum_{i}^{6} \pi_i = 1. \tag{6.16}$$

In the upcoming section, the operator profit aspects towards designing a green communication system is analyzed.

6.4 Operator Profit Analysis and Sustainable System Design

This section first presents lower bounds for CAPEX provisioning at hMBS. Then the system is analyzed from a service provider's perspective, analyzing the various cost parameters involved and aim to design a green network.

6.4.1 HAPS CAPEX Provisioning for Sustainable RAN

The current literature assumes that the HAPS is sufficiently powered with renewable energy sources for its sustenance in stratosphere as well as for RAN functioning [144]. Since the hMBS deployment involves CAPEX (including solar panels and storage batteries) to the telecom provider, thus it is not practical to consider user offloading to hMBS free of cost.

The hMBS is equipped with a uniform planar antenna array to mitigate the free space path loss. The energy consumption by the hMBS is assumed to be similar to a tMBS as proposed in [144]. Accordingly, $E_H(t) = N^H_{TRX}(\mathcal{P}_0 + \theta_1 P^T_H)$ denotes the temporal energy consumption by a hMBS when radiating P^T_H power through N^H_{TRX} antenna elements. For sustainable RAN, the hMBS should be equipped with sufficient CAPEX such that on an average the daily energy harvest at the hMBS (for RAN) be greater than the hMBS RAN energy consumption. The lower bounds for minimum CAPEX provisioning in a hMBS are,

$$\begin{aligned} N^H_{PV} &\geq \frac{\sum_t \mathbb{E}\left(E_H(t)\right)}{\eta \times \sum_t \mathbb{E}\left(H_{1\text{KW}}(t)\right)} \\ N^H_B &\geq \frac{\left(\sum_t H_{N_{PV}}(t) - \sum_t E_H(t)\right)}{\delta \times \beta^{cr}}. \end{aligned} \tag{6.17}$$

Here, $H_{1\text{KW}}$ denotes the energy harvested by a 1 KW (i.e., unit rated) PV panel and η denotes the efficiency of the PV panels. It may be noted that the CAPEX provisioning at a hMBS is a function of the number of antenna elements in the array equipped with hMBS to mitigate the free space path loss, and is computed as

$$CAPEX^H = C_{PV} N^H_{PV} / \mathcal{L}_{PV} + C_B N^H_B / \mathcal{L}_B. \tag{6.18}$$

Here, $C_{PV}, \mathcal{L}_{PV}$ and $C_B, \mathcal{L}_B$ denote the cost and lifetime of unit PV panel and battery, respectively.

6.4.2 Energy Sustainable System Design

The main objective of this work is to use the grid infrastructure only for selling energy back to the grid and not for grid energy procurement. The probability of selling excess energy and offloading excess users to the hMBS is given as

$$\begin{aligned} \mathbb{P}_{sell} &= \pi_{QS,S} + \pi_{QV,S} \\ \mathbb{P}_{offload} &= \pi_{QS,D} + \pi_{QV,D} + \pi_{QV,I} + \pi_{QV,S}. \end{aligned} \tag{6.19}$$

The network aims to maximize the network user service by ensuring that the QoS of all temporally active users can be fulfilled. Accordingly the following rate maximization problem is formulated.

$$\begin{aligned} \mathcal{P}1.\quad & \max_{\pi_{ub},\pi_{uh},\mathbf{P}} (\pi_{ub}(t)r_{ub}(t) + \pi_{uh}(t)r_{uh}(t))\ \forall u \in \mathbf{U} \\ & \text{such that, } \pi_{ub}, \pi_{uh} \in \{0, 1\} \\ & \qquad r_{ub}(t), r_{uh}(t) \geq r_{th} \\ & \qquad \pi_{ub}(t) = 1 - \pi_{uh}(t) \end{aligned} \tag{6.20}$$

It may be noted that π_{uh} denotes the association of a user with the hMBS. Since the above problem is a combinatorial mixed-integer non-linear problem, it is NP hard in nature. It is solved by Algorithm 2 having complexity $\mathbb{O}(T \times U)$. For a general network having B tMBSs, the complexity will be $\mathbb{O}(T \times U \times B)$. Algorithm 2 outputs the number of user served in the network N. The operator profit is computed considering the following factors: revenue earned by serving users R_{ser}, revenue earned by selling excess energy to the grid R_{sell}, and capital expenditure $CAPEX$. It may be noted that in order to realize green network, the tMBS does not procure energy from grid, rather offloads the user to hMBS. The offloading at hMBS is not free to the operator. The operator incurs CAPEX in solar provisioning the RAN at hMBS as discussed in Section IV-A. Net profit $\mathcal{P}_1 = R_{ser} + R_{sell} - CAPEX^H - CAPEX^B$. Here, $CAPEX^B = (C_{PV}N_{PV}/\mathcal{L}_{PV} + C_B\eta_B/\mathcal{L}_B)$ denote the CAPEX incurred in solar provisioning tMBS. Revenue earned by serving users is computed as $R_{ser} = C_{ser}\sum_t N(t)$, while revenue earned by selling energy is computed as $R_{sell} = \sum_t (B_b'(t) - \beta^{max})$ in the event of battery overflow.

As a competitive approach, in case the operator doesn't wish to offload users to hMBS, and resorts to purchasing energy from the grid, the net operator profit will be computed as $\mathcal{P}_2 = R_{ser} + R_{sell} - CAPEX' - C_{buy}$. Here $C_{buy} = C_b \sum_t (\beta_{cr} - B_b'(t))$ with $CAPEX' \neq CAPEX^B$. In the upcoming section, the results and inferences are discussed.

6.5 Results and Discussion

In this section, the simulated results along with key observations and inferences, generated through MATLAB software are discussed. The values of the simulation parameters considered are provided in Table 6.1. The entire analysis is conducted for an annual time frame, $T = 8760$ Hr.

Through Fig. 6.3a, the influence of QoS guarantee on the number of users being served in the network is illustrated. Figure 6.3a illustrates the variation of tMBS power level (in Watts) with increasing user density. Further, this variation is studied at multiple user QoS values. It is observed that with increasing user QoS, the tMBS is able to serve fewer network users, with the corresponding tMBS power level

Algorithm 2: User association and offloading

Result: $N = \sum_u \pi_{ub} + \sum_u \pi_{uh}$, **P**
Input: **U**, $\rho(t)$, $\mathbf{g}(t)$, W, f, σ^2, $\mathcal{P}_{max}$
Initialize: $N = 0$
for $t = \{1, \ldots, T\}$ **do**
  for $u = \{1, \ldots, U(t)\}$ **do**
    $d_{ub} \longleftarrow \sqrt{(x_u - x_{bs})^2 + (y_u - y_{bs})^2}$
    Compute $SINR_{ub}$ from (1)
    $r_{ub}(t) \longleftarrow W_{ub} \log_2(1 + SINR_{ub}(t))$
    if $(r_{ub}(t) \geq r_{th})$ **then**
      $N \longleftarrow N + 1$
      $\pi_{ub} \longleftarrow 1$
      Compute $\mathcal{P}_{ub}$, P_b^T from (6.11)
    end
    else
      Compute P_r from hMBS using (1)–(8)
      Compute $r_{uh} = f \log_2(1 + SINR_{uh})$
      if $(r_{uh}(t) \geq r_{th})$ **then**
        $\pi_{uh} \longleftarrow 1$, $N \longleftarrow N + 1$
      else
        user remains unserved
      end
    end
  end
end
end

Table 6.1 Parameter values used in simulations

Parameter	Value	Parameter	Value
$\mathcal{A}$	1 km^2	H_{hap}	20 km
W	20 MHz	R_e	6378 km
β^{cr}	2460 Wh	f_c	2.5 GHz
σ^2	−150 dBm/Hz	CL_{NLoS}	25.5 dB
$\mathcal{P}_{max}$	40 W	CL_{LoS}	9.2 dB
$\mathcal{P}_{uH}^T$	49 dBm	X_{LoS}	$\mathcal{N}(0, \sigma_{LoS})$
θ	90°	X_{NLoS}	$\mathcal{N}(0, \sigma_{NLoS})$
σ_{NLoS}	0.6 dB	σ_{LoS}	1.2 dB
N_{TRX}^H	32 elements (8 dBi each)	η	0.5
C_{PV}	1300$	C_B	216$

saturating at 40 W, which is the FCC mandated power level for a tMBS [109]. In this work the minimum QoS requirement considered, $r_{th} = 1$ Mbps.

Figure 6.3b–c illustrate the variation of network QoS. It may be noted that network QoS is measured with respect to the percentage of users served by the network, especially during the peak hours (9 AM–6 PM) in accordance with the traffic profile considered in Fig. 6.1b. It is observed from Fig. 6.3b that the tMBS

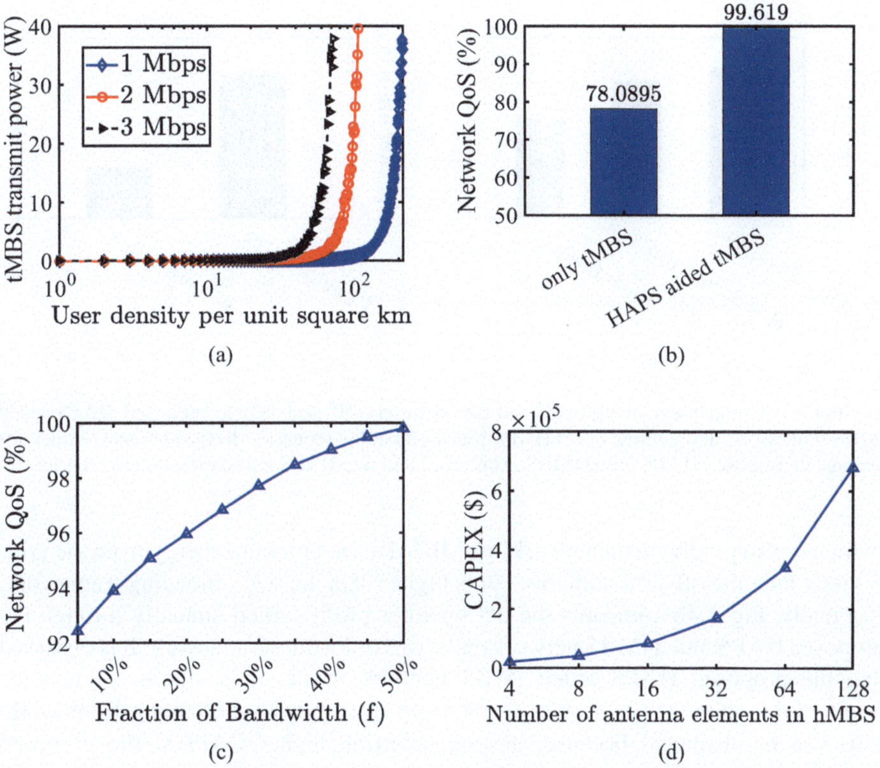

Fig. 6.3 (**a**) Effect of minimum QoS on variation of tMBS transmit power level (in Watts) with user density, (**b**) Network QoS variation at peak hours (9 AM–6 PM) in only tMBS scenario (no offloading) and HAPS aided tMBS scenario (hMBS operating at 50% capacity), (**c**) Network QoS variation in peak hours with limited hMBS capacity in HAPS aided tMBS network, (**d**) Variation of hMBS CAPEX for RAN with number of antenna elements equipped with hMBS

alone is able to serve only about 78% of the users, while a hMBS (operating at 50% frequency) aided tMBS network provides a gain of roughly 21.6%, serving close to 99.6% users. On varying the capacity of hMBS available to serve the considered close area (as in Fig. 6.3c), it is observed that the hMBS aided tMBS network still performs better than a purely tMBS network. Figure 6.3d illustrates the solar CAPEX provisioning required in a hMBS for RAN functioning as a function of number of antenna elements equipped with the hMBS. It is observed that the CAPEX of hMBS increases exponentially with increasing number of antenna elements.

Through Fig. 6.4a the CAPEX incurred to the mobile operator is illustrated. As a benchmark, an off-grid standalone tMBS framework [69] and a without (w/o) offloading framework wherein the tMBS has flexibility of grid energy procurement in addition to energy selling, have been considered for comparison purpose. It is observed that the CAPEX incurred in the proposed HAPS aided tMBS network

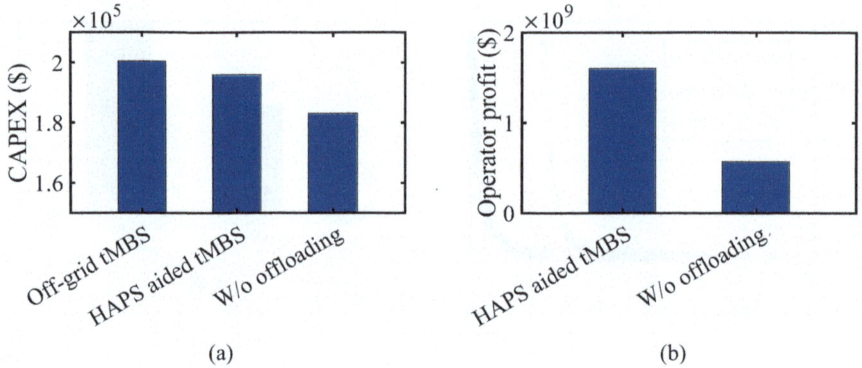

Fig. 6.4 (**a**) Comparison of optimal CAPEX in purely off-grid tMBS, proposed HAPS aided tMBS framework, and without (w/o) HAPS based offloading scenario, (**b**) Comparison of operator revenue in proposed HAPS aided tMBS framework and w/o HAPS based offloading scenario

(which is also purely green network, as tMBS does not procure energy from the grid) is lesser than the off-grid scenario, but is higher than the w/o offloading framework.

Finally, Fig. 6.4b compares the net operator profit earned annually through the proposed HAPS aided tMBS network and a w/o offloading framework. It is observed that the proposed HAPS aided tMBS network results in a significant revenue profit to the operator (gain close to 64.51%) over the w/o offloading framework. This can be attributed because, despite incurring higher CAPEX, the proposed framework obtains higher revenue by user service (higher QoS service as shown in Fig. 6.3b). Further, there is no grid energy procurement in the proposed HAPS aided tMBS network, resulting in reduced operational expenditure to the operator, thus consolidating the net profit earned.

6.6 Summary

This chapter has proposed an analytical framework of a hMBS offloading aided and smart grid connected green communication network. To realize green network, the framework has aimed to limit the flexibility of tMBS to procure energy from grid, rather has used grid connectivity purely for energy selling purpose. The analytical framework has been modeled as a six state DTMC by studying the communication and energy aspects in the considered system. The proposed HAPS aided and grid connected tMBS network has been observed to provide significant gains in user QoS and operator profit despite incurring higher CAPEX. The proposed framework is expected to incentivize mobile operators, achieve grid energy independence, and pave the way towards green communication systems.

Chapter 7
Extending Energy Balancing to Consumer Utility Networks

7.1 Introduction

In this chapter, we aim to extend the concept of energy balancing in telecommunication network to a general consumer utility network. The idea is to showcase the importance of the core concept of energy balancing in energy aware network design.

Traditional residential systems have been powered through the power grid (also termed as marcogrid, in context of this chapter). The load consumption patterns in residential systems vary with user depending on the time of day. In residential societies, where user load behavior exhibit a higher degree of correlation, the load consumption is generally seen much higher in peak hours, and much lower in other parts of the day. High demand of energy from the grid during peak hours, often result in high stress on the grid itself, making the grid prone to breakdown or vulnerable to fault leading to energy outages. Load balancing strategy has been a go to approach in residential networks, especially in such peak hours. The grid energy purchase price also becomes higher during peak hours, to incentivize users from reducing the load during peak hours [145]. Load balancing frameworks includes delaying the service by a user in peak hours, and executing those services at other hours. It is notable, that a user "quality of service" gets affected because of load balancing, as a user is unable to execute the service when required and needs to wait.

Energy balancing can greatly assist in mitigating the reduction of user experience, due to reduction of QoS. Residences can be networked using a micro-grid. Such microgrid networked systems, can cooperatively engage in energy balancing among the networked nodes, without relying on the marcogrid for power. Hence, energy balancing not only reduces the load on the marcogrid, but also improves the user QoE greatly.

Traditional macrogrid connected residential systems are powered mainly by carbon emitting thermal plants. The US Energy Information Administration (EIA) estimates around 21% increase in residential electricity consumption by 2040 [146]. While meeting energy efficiency is crucial to reduce the carbon footprint, cost

L.-C. Wang et al., *Green Energy Driven Integrated Smart Grid and Wireless Networks*, Wireless Networks, https://doi.org/10.1007/978-3-032-20509-4_7

has emerged as an important factor towards realizing the energy efficient solutions [147]. Hence there has been a significant need to develop scalable, energy efficient, and cost optimal solutions for residential frameworks. In this work, we present a futuristic energy cooperative framework for realizing profitable and self-sustainable green residential systems.

7.1.1 State-of-the-Art and Motivation

Traditional strategies to reduce carbon footprint have been to optimize the power consumption of the macrogrid connected residences [148, 149]. The framework in [148] presented a peak load shifting strategy in accordance with real-time energy prices. Discrete time Markov model based approaches were proposed in [83, 149]. Both the proposed frameworks consider different Markov models for the energy harvest, load, and battery. The combined Markov model is computationally intensive and comprises of a large number of states. Recently, the authors in [145] presented a game theory based grid connected residential framework on gauging the demand response based on time varying energy prices. Along this line, the framework in [150] presented a decentralized framework for load management in smart grids.

Provisioning residential systems with solar energy supplies in addition to macrogrid connectivity has been an attractive solution [148, 151]. While these dual-powered residential systems are not carbon free, achieving self-sustainability is impractical, incurring a significant capital expenditure (CAPEX) to the residence owner [152]. Additionally, in a multi-residential system, being a non-cooperative framework, the spatio-temporal load variations may often result in improper utilization of the green energy with some residences [48].

As an advance, in this chapter, we present a Markovian analysis approach to investigate the benefit of energy cooperative residential systems.

7.1.2 Key Chapter Insights

The key insights from this chapter are as follows.

- This chapter extends the concept of energy balancing to ambient powered, energy aware residential networks. From a system perspective, we first analyze a macro grid connected and ambient powered residential system. The analysis is performed through a Markovian framework. Performance evaluation is performed through the optimal CAPEX provisioning required per household.
- Next, we present the analysis of a cooperative residential system. The concept of energy balancing is re-examined among the micro-grid networked and solar powered residences. The analysis is performed through a Markovian framework, towards designing green residential system.

- The analysis showcases the energy management, including optimal amount of energies sold, purchased or to be shared among the residences. The concept is expected to be useful in incentivizing residential users towards energy balancing, and thus offsetting peak grid hours, thereby assisting in increased revenue to the grid users.

7.2 Solar Powered Residential System Model

We consider a residential society having $\mathcal{R}$ dual-powered, solar-enabled and macrogrid connected, non-cooperative residences represented as $\mathbf{R} = \{1, \cdots, r, \cdots, \mathcal{R}\}$. Each residence r generates $L_r(t)$ demand hourly and depends on the solar harvest and/or the macrogrid to meet the demand as shown in Fig. 7.1a. We discuss the load and energy harvest profile in the upcoming subsections.

7.2.1 *Residential Load Profile*

The average hourly load, $L(t)$, generated by a residence is illustrated in Fig. 7.1b [153]. We generate temporal load variations of varying skewness [49] between the residences as explained below.

$$\begin{aligned} \psi_r(t) &= \frac{e^{Sr}}{\sum_{r=1}^{\mathcal{R}} e^{Sr}}, \\ \text{s.t.,} \sum_{r=1}^{\mathcal{R}} \psi_r(t) &= \sum_{r=1}^{\mathcal{R}} \frac{e^{Sr}}{\sum_{r=1}^{\mathcal{R}} e^{Sr}} = 1. \end{aligned} \tag{7.1}$$

Here, $\psi_r(t)$ represents the skewed fraction of average load that a residence will generate and S represents the *skewness parameter* influencing the temporal variations among the residence loads. Figure 7.1c illustrates the temporal load variations depending on S, for a three residence cluster. For a multi-residence system, depending on S, $L_r(t) = L(t) \times \psi_r(t) \times \mathcal{R}$.

7.2.2 *Energy Harvest Profile*

Each residence r is assumed to be enabled with N_{PV} panels of rating R_{PV} and N_B batteries of capacity B_{cap}, harnessing $H_r(t) = I_r(t)\eta N_{PV}$ Watts of energy hourly. Here, $I_r(t)$ represents the hourly solar irradiance obtained from National Renewable Energy Laboratory (NREL) [112] and η represents the PV panel efficiency. The

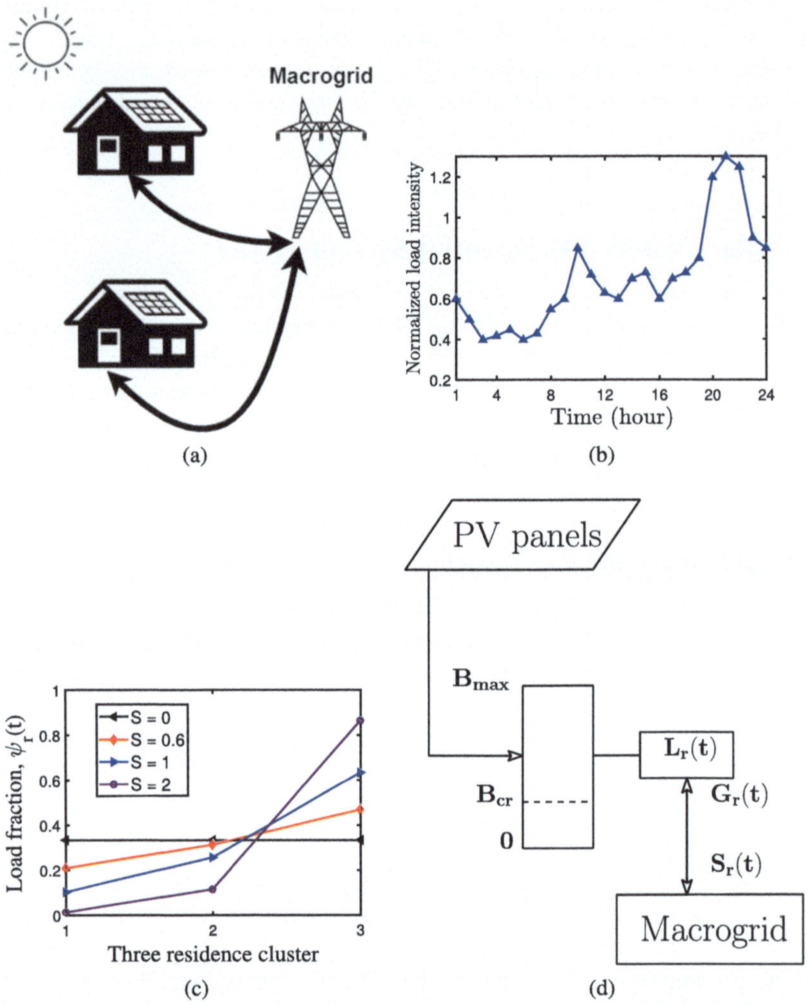

Fig. 7.1 (**a**) Pictorial representation of a dual-powered non-cooperative MA-RES framework, (**b**) Normalized hourly load profile, (**c**) Illustration of skewed load generation for a three residence cluster, (**d**) Energy flow in a MA-RES framework

upper limit of the energy storage is given as $\beta_{max} = N_B B_{cap}$ and the critical threshold level below which energy cannot be disbursed to the residence is given as $\beta_{cr} = \delta\beta_{max}$. Here, δ represents the depth of discharge fixed in the battery management system. The hourly battery level of a residence r is computed as

$$\begin{aligned} \beta_r'(t) &= \beta_r(t-1) + H_r(t) - L_r(t) \\ \beta_r(t) &= \min\{\max\{\beta_r'(t), \beta_{cr}\}, \beta_{max}\}. \end{aligned} \tag{7.2}$$

In the upcoming section, we discuss the optimal design framework for a MA-RES system.

7.3 Optimal Macrogrid Connected System Design

In this section, we discuss the optimal design of a MA-RES framework, through a three state discrete time Markov approach to characterize the battery storage of a residence.

7.3.1 *Markovian Modeling of Energy Storage*

Depending on the hourly residential load and energy harnessed by the PV panels, the residential battery level can transit among three states, namely, deficit (D), intermediate (I), and surplus (S). The hourly deficit $D_r(t)$ and surplus $S_r(t)$ energies in a residence r are computed as

$$\begin{aligned} D_r(t) &= |\min\left\{0, B_r'(t) - \beta_{cr}\right\}| \\ S_r(t) &= \max\left\{0, B_r'(t) - \beta_{max}\right\}. \end{aligned} \tag{7.3}$$

The energy flow in a MA-RES system is shown in Fig. 7.1d. The deficit and the surplus states refer to the condition when the battery level goes below the critical level ($\beta_r'(t) < \beta_{cr}$) and above the battery upper limit ($\beta_r' > \beta_{max}$), respectively. The intermediate state refers to the condition when the battery level is between β_{max} and β_{cr}. This transition is modeled via a three state Markov model as shown in Fig. 7.2. The transition matrix ($\underline{T}_1$) corresponding to the Markov model is represented below, with the transition probabilities from state $i \longrightarrow j$ being $P_{ij}\ \forall\, i, j \in \{D, I, S\}$.

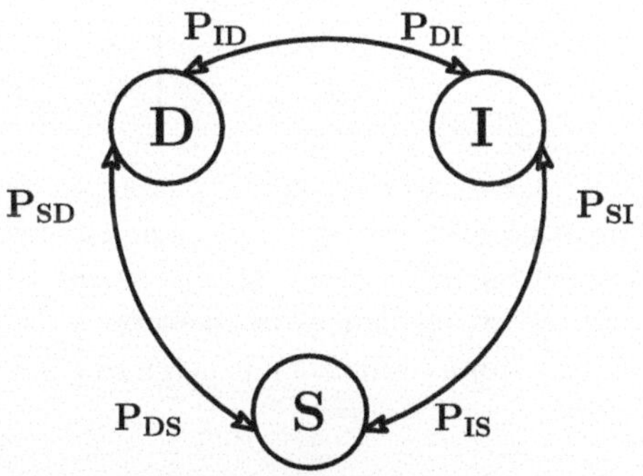

Fig. 7.2 Proposed Markov model for a MA-RES system

$$\underline{\mathrm{T}}_1 = \begin{bmatrix} P_{DD} & P_{DI} & P_{DS} \\ P_{ID} & P_{II} & P_{IS} \\ P_{SD} & P_{SI} & P_{SS} \end{bmatrix}. \tag{7.4}$$

The steady state probabilities of the three states are given as $\underline{\pi}_1 = \{\pi_D, \pi_I, \pi_S\}$, respectively. The matrices, $\underline{\pi}_1$ and $\underline{\mathrm{T}}_1$, are solved using the system of equations given below to obtain the steady state probabilities in terms of the transition probabilities.

$$\begin{aligned} \underline{\pi}_1 &= \underline{\pi}_1^T \underline{\mathrm{T}}_1 \\ \sum_i \pi_i &= 1 \,\forall i \in \{D, I, S\}. \end{aligned} \tag{7.5}$$

It may be noted that the steady state probabilities, π_D and π_S, represent the probabilities of energy outage and energy overflow, respectively.

7.3.2 *Achieving Cost Effectiveness and Sustainability*

In this subsection, we formulate an optimization problem towards computing the optimal CAPEX such that the MA-RES system becomes profitable and further self-sustainable. The grid energy procurement occurring in the event of transition to a deficit state is given as

$$G_r(t) = \begin{cases} 0, & \beta_r'(t) \geq \beta_{cr} \\ |\beta_r'(t) - \beta_{cr}|, & \beta_r'(t) < \beta_{cr}. \end{cases} \tag{7.6}$$

Thus, $G_r(t) = \sum_t D_r(t)$. The hourly operational revenue, $\mathrm{OP}_{rev1}(t)$, earned by a residence r is given as

$$\mathrm{OP}_{rev1} = \begin{cases} C_{Sell} S_r(t), & \text{if } \beta_r'(t) > \beta_{max} \\ C_{buy} G_r(t), & \text{if } \beta_r'(t) < \beta_{cr}. \end{cases} \tag{7.7}$$

Here, C_{Sell} and C_{buy} correspond to the price of selling/buying unit energy to/from the macrogrid. Hence, it can be noted that the energy selling or energy procurement by a residence r from the macrogrid are disjoint with each other. Let t_1 and t_2 denote the time indices when energy selling/buying occurs. Then the net operational revenue for a residence r is given as

$$\begin{aligned}\text{OP}_{rev1} &= \sum_{t_1} C_{Sell} S_r(t_1) - \sum_{t_2} C_{buy} G_r(t_2) \\ &= \sum_{t_1} C_{Sell} \left(\beta_r'(t_1) - N_B B_{cap}\right) - \sum_{t_2} C_{buy} \left(\delta N_B B_{cap} - \beta_r'(t_2)\right) \\ &= f\left(N_{PV}, N_B\right).\end{aligned} \tag{7.8}$$

Here, T_1 and T_2 represent the cardinality of sets indexed by t_1 and t_2. The net profit $\mathcal{P}$ earned by a residence is then computed as

$$\mathcal{P} = \text{OP}_{rev1} - CAPEX, \tag{7.9}$$

where $CAPEX = (N_{PV} C_{PV}/L_{PV} + N_B C_B/L_B)$, with C_{PV}, C_B being the costs of unit PV panel and battery and L_{PV}, L_B being the corresponding lifetime. An optimal CAPEX computation problem such that a residence incurs net profit over a time duration is formulated as

$$\begin{aligned} \min_{N_{PV}, N_B} \quad & CAPEX = (N_{PV} C_{PV}/L_{PV} + N_B C_B/L_B) \\ \text{s.t.,} \quad & \text{OP}_{rev1} \geq 0, N_{PV} > 0, N_B > 0. \end{aligned} \tag{7.10}$$

Further, we also formulate a problem to design a self-sustainable system as given below

$$\begin{aligned} \min_{N_{PV}, N_B} \quad & (N_{PV} C_{PV}/L_{PV} + N_B C_B/L_B) \\ \text{s.t.,} \quad & \pi_D = 0, N_{PV} > 0, N_B > 0. \end{aligned} \tag{7.11}$$

In the upcoming section, we propose a multi-residence cooperative energy transfer framework to improve upon the CAPEX incurred per residence for achieving self-sustainability.

7.4 Cooperative Multi-residence System Design

In this section, we propose a microgrid based cooperative energy transfer mechanism among multiple dual-powered residential systems, called MI-RES, as shown in Fig. 7.3a. The microgrid is visualized to be a local co-located energy reserve at the community/society level (if the residences are flats in an apartment) or is a virtual entity controlling energy flow among the individual solar-enabled residences.

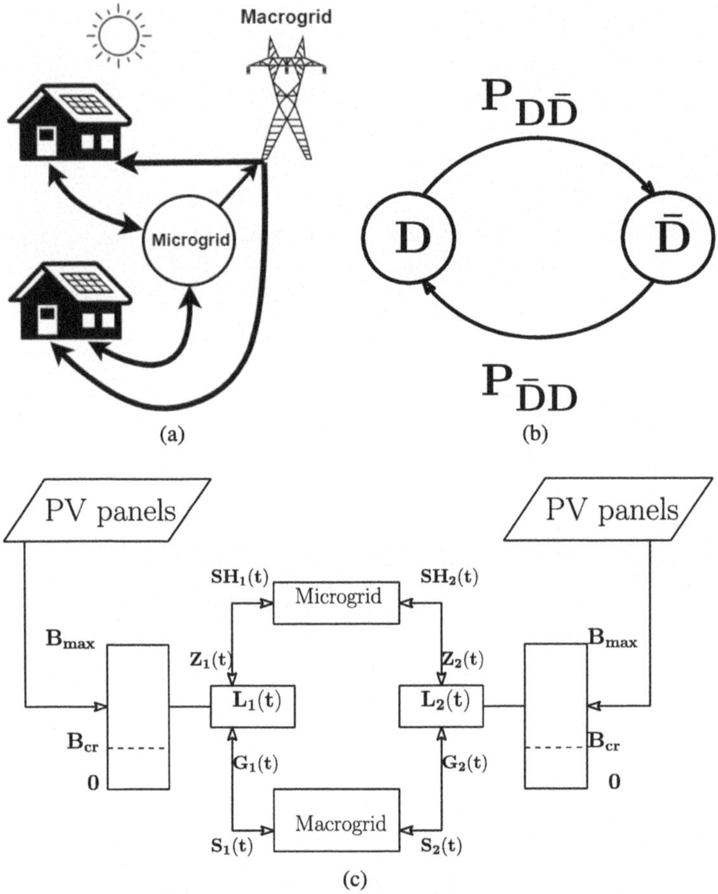

Fig. 7.3 (**a**) Pictorial representation of the proposed MI-RES framework, (**b**) Corresponding Markov model for battery storage, (**c**) Energy flow in the MI-RES framework

7.4.1 Markovian Characterization of Energy Storage

The battery level of a MI-RES system is proposed to be in two states namely, deficit (D) and non-deficit ($\bar{D}$), as illustrated in Fig. 7.3b. The corresponding transition probabilities are shown in the transition matrix $\underline{T}_2$ as, $P_{ij}\ \forall\ i, j \in \{D, \bar{D}\}$.

$$\underline{T}_2 = \begin{bmatrix} P_{DD} & P_{D\bar{D}} \\ P_{\bar{D}D} & P_{\bar{D}\bar{D}} \end{bmatrix}. \tag{7.12}$$

The steady state probabilities for the two state Markov chain are given by $\underline{\pi}_2 = \{\pi_D, \pi_{\bar{D}}\}$. The steady state values are obtained by solving the system of equations given below.

$$\underline{\pi}_2 = \underline{\pi}_2^T \underline{\text{T}}_2$$
$$\sum_i \pi_i = 1 \,\forall\, i \in \{D, \bar{D}\}. \tag{7.13}$$

It may be noted that π_D here corresponds to the probability of energy outage. Let, at an hour t, R residences out of $\mathcal{R}$ be in state D (indexed by r), while the remaining R' residences be in state $\bar{D}$ (indexed by r'), such that $R + R' = \mathcal{R}$.

The energy flow occurring in a MI-RES framework is illustrated in Fig. 7.3c. When the battery level is in state D, the residence can supplement the deficit energy ($D_r(t) = |\min\{0, \beta_r'(t) - \beta_{cr}\}| = Z_r(t) + G_r(t)$) either from the microgrid or the macrogrid. Here, $Z_r(t)$ refers to the green energy transferred, to a residence r in state D, cooperatively by the residences in state $\bar{D}$ and $G_r(t)$ refers to the energy purchased by a residence r from the macrogrid. When the battery level is in state $\bar{D}$, the residence after meeting its demand can share $SH_{r'}(t) = \max\{0, \beta_r'(t) - \beta_{cr}\}$ energy to the microgrid. Thus, depending on the hourly energy harvest and the load of each residence, a society will have $\sum_{r'} SH_{r'}(t)$ sharable energy and $\sum_r D_r(t)$ deficit energy. The following cases may arise in a multi-residence MI-RES framework.

$$\text{If} \sum_{r'} SH_{r'}(t) \geq \sum_r D_r(t), \text{then:} \begin{cases} \sum_r G_r(t) \overset{(a)}{=} 0 \\ \sum_r Z_r(t) \overset{(b)}{=} \sum_r D_r(t) \\ \sum_{r'} S_{r'}(t) \overset{(c)}{=} \\ \sum_{r'} SH_{r'}(t) - \sum_r D_r(t). \end{cases} \tag{7.14}$$

$$\text{If} \sum_{r'} SH_{r'}(t) < \sum_r D_r(t), \text{then:} \begin{cases} \sum_r G_r(t) \overset{(a)}{=} \\ \sum_r D_r(t) - \sum_{r'} SH_{r'}(t) \\ \sum_r Z_r(t) \overset{(b)}{=} \sum_{r'} SH_{r'}(t) \\ \sum_{r'} S_{r'}(t) \overset{(c)}{=} 0. \end{cases} \tag{7.15}$$

From (14b) and (15b), we can rewrite (14c) and (15a) as $\sum_{r'} S_{r'}(t) = \sum_{r'} SH_{r'}(t) - \sum_r Z_r(t)$ and $\sum_r G_r(t) = \sum_r D_r(t) - \sum_r Z_r(t)$, respectively. In the next subsection we will discuss about computing the optimal CAPEX in a MI-RES framework, to achieve self-sustainability.

7.4.2 Optimal CAPEX Towards Achieving Self-sustainability

In this subsection, we formulate an optimization problem towards computing the optimal CAPEX in a multi-residence cooperative cluster towards achieving self-sustainability. The net operational revenue ($\mathrm{OP}_{rev2}(t)$) earned by the society at an hour is given by

$$\mathrm{OP}_{rev2}(t) = \underbrace{C_{Sell}\sum_{r'} S_{r'}(t)}_{\text{Selling}} - \underbrace{C_{buy}\sum_{r} G_r(t)}_{\text{Buying}} - \underbrace{C_{sh}\sum_{r} Z_r(t)}_{\text{Transfer}}. \tag{7.16}$$

The above equation can be further expanded and simplified as

$$\begin{aligned} \mathrm{OP}_{rev2} &= C_{Sell}\left(\sum_{r'} SH_{r'}(t) - \sum_{r} Z_r(t)\right) - C_{sh}\sum_{r} Z_r(t) \\ &\quad - C_{buy}\left(\sum_{r} D_r(t) - \sum_{r} Z_r(t)\right) \\ &= \left(C_{buy} - C_{Sell} - C_{sh}\right)\sum_{r} Z_r(t) + C_{Sell}\sum_{r'} SH_{r'}(t) \\ &\quad - C_{buy}\sum_{r} D_r(t). \end{aligned} \tag{7.17}$$

From (14b) and (15b), we observe that

$$\sum_{r} Z_r(t) = \min\left\{\sum_{r} D_r(t), \sum_{r'} SH_{r'}(t)\right\}. \tag{7.18}$$

Therefore, $\mathrm{OP}_{rev2}(t)$ can be expressed as

$$\mathrm{OP}_{rev2}(t) = \begin{cases} (-C_{Sell} - C_{sh})\sum_r D_r(t) + C_{Sell}\sum_{r'} SH_{r'}(t), \\ \qquad \text{if } \sum_r D_r(t) \le \sum_{r'} SH_{r'}(t). \\ (C_{buy} - C_{sh})\sum_{r'} SH_{r'}(t) - C_{buy}\sum_r D_r(t), \\ \qquad \text{if } \sum_r D_r(t) > \sum_{r'} SH_{r'}(t). \end{cases} \tag{7.19}$$

Hence we observe that, if $\sum_r D_r(t) \le \sum_{r'} SH_{r'}(t)$, then the operational revenue at that hour does not depend on C_{buy} (i.e., intuitively no macrogrid energy procurement). Also, if $\sum_r D_r(t) > \sum_{r'} SH_{r'}(t)$, then the hourly operational revenue does not involve C_{Sell} (i.e., no energy selling involved to earn additional revenue). From (7.19) we observe that $0 \le C_{sh} << C_{Sell} << C_{buy}$, such that $\mathrm{OP}_{rev2}(t) \ge 0$ if $\sum_r D_r(t) \le \sum_{r'} SH_{r'}(t)$ and $\mathrm{OP}_{rev2}(t) \le 0$ if $\sum_r D_r(t) >$

$\sum_{r'} SH_{r'}(t)$. If $C_{sh} = 0$, we term the MI-RES framework as *social cooperation*, and if $0 < C_{sh} \leq C_{buy}$ then it is termed as *business cooperation*. It may be noted that if $C_{sh} > C_{buy}$ then there is no point of cooperation. The net profit earned in the MI-RES framework can be represented as

$$\mathcal{P} = \sum_{t} \text{OP}_{rev2}(t) - N_{PV}C_{PV}/L_{PV} - N_B C_B/L_B. \tag{7.20}$$

Thus, the problem to design a cooperative, macrogrid energy independent system, is formulated as

$$\min_{N_{PV}, N_B} CAPEX = \sum_{r}^{\mathcal{R}} (N_{PV}C_{PV}/L_{PV} + C_B N_B/L_B) \tag{7.21}$$

$$\text{s.t., } N_{PV} \geq 0, N_B \geq 0, \mathcal{R} \geq 2, \pi_D = 0.$$

Thus, the formulated problem in a MI-RES framework depends on the solar provisioning (N_{PV}, N_B) and the number of cooperative residences ($\mathcal{R}$), or the residence cluster size. $\mathcal{R} \geq 2$ since at least two residences are required for cooperation. The formulated problems in (7.10), (7.11), and (7.21) are observed to be mixed-integer linear optimization problems. These problems have been solved using standard linear optimization solvers in MATLAB. We will discuss the simulated results obtained and our inferences and observations in the upcoming section.

7.5 Results and Discussion

The simulations have been performed in MATLAB R2020a software, with annual hourly solar irradiance data of Jaipur city in India. The parameter values used in simulations are, $\delta = 0.3$, $\eta = 0.5$, $B_{cap} = 2460$ Wh, $R_{PV} = 1$ KW, $C_{buy} = 0.08$ USD, $C_{Sell} = 0.057$ USD, $C_{sh} = 0.02$ USD, $C_{PV} = 1500$ USD, $C_B = 400$ USD, $L_B = 0.25$, $L_{PV} = 1$. Through our results, we first illustrate the optimal CAPEX required to design a profitable and further self-sustainable MA-RES framework. Then we illustrate the working of the proposed MI-RES framework demonstrating the reduction in CAPEX per residence with the proposed framework.

Through Fig. 7.4a and b we illustrate the design of a profitable and self-sustainable MA-RES framework. Figure 7.4a shows the variation of annual profit with change in solar provisioning, i.e., N_{PV} and N_B. It is observed that the MA-RES framework becomes profitable, i.e., the solar provisioning at which annual profit is greater than 0, at $(N_{PV}, N_B) = (8, 13)$. The corresponding CAPEX is computed to be 32,800 USD.

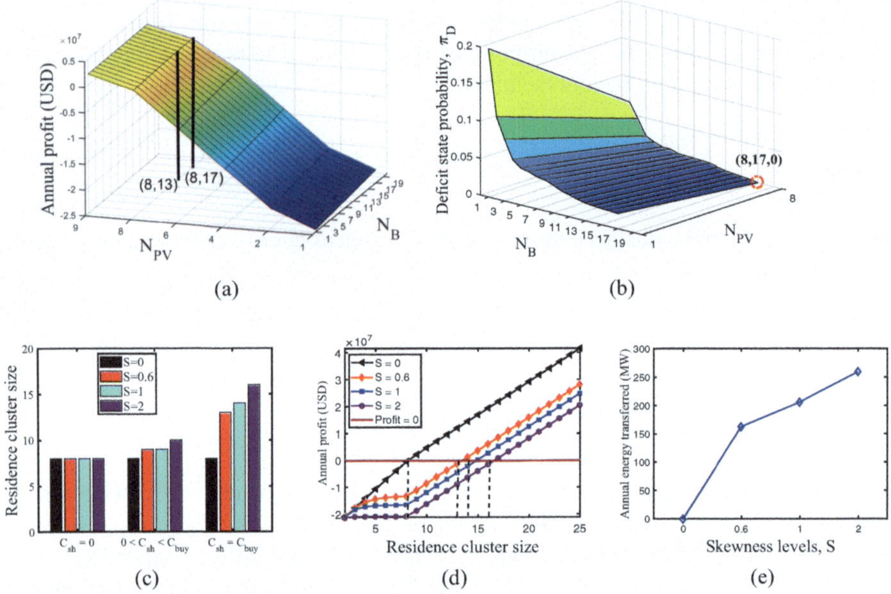

Fig. 7.4 (**a**) Annual optimal profit in a single residence MA-RES framework, (**b**) Steady state probability variation of deficit state, π_D, in a MA-RES framework, (**c**) Variation of cluster size with C_{sh} to achieve profitability in a MI-RES system, with $(N_{PV} = 1, N_B = 1)$, (**d**) Variation of annual profit with residence cluster size at $C_{sh} = C_{buy}$, (**e**) Maximum annual energy transferred to a residence in the MI-RES system, achieving self-sustainability when $(N_{PV} = 1, N_B = 1)$ (corresponding cluster size = 10)

Figure 7.4b represents the variation in steady state probability of the deficit state (π_D) with a change in solar provisioning. *A general trend inferred from Fig. 7.4b is that, π_D decreases with an increase in N_B. It is also inferred that with an increase in N_{PV}, the rate of decrease of π_D with increasing N_B, decreases.* It is observed from Fig. 7.4b that the steady state probability of the deficit state (π_D) becomes zero at $(N_{PV}, N_B) = (8, 17)$, corresponding to a CAPEX of 39,200 USD. Since π_D represents the long-term energy outage probability, hence $\pi_D = 0$ corresponds to a scenario where there is no transition into state D in the Markov model proposed in Sect. 7.3.1 and hence we can say that the MA-RES system becomes self-sustainable at that (N_{PV}, N_B).

Next, we demonstrate the working of a cooperative MI-RES system, through Fig. 7.4c–e. Each residence of the cooperative framework is solar provisioned as $(N_{PV} = 1, N_B = 1)$. Through Fig. 7.4c and d we demonstrate the variation of cluster size with changing values of C_{sh}, to achieve a cost profitable MI-RES system. It can be observed from Fig. 7.4c that $C_{sh} = 0$ (i.e., social cooperation) results in all the skewness levels attaining profitability at the same cluster size. Upon increasing C_{sh} such that, $0 < C_{sh} \leq C_{buy}$, it is observed that *higher skewness levels require much higher collaborative residences in order to become profitable.*

Table 7.1 Cluster size versus CAPEX in MI-RES system, for self-sustainability

(N_{PV}, N_B)	(1, 1)	(2, 1)	(3, 1)	(4, 1)	(5, 1)
Cluster size	10	5	4	3	2
(N_{PV}, N_B)	(1, 2)	(1, 5)	(1, 10)	(1, 15)	(1, 20)
Cluster size	10	10	9	9	9

For instance, when $C_{sh} = C_{buy}$, at $S = 2$ (extreme skewness level), a MI-RES residential setting requires a cooperative residence cluster of 16 residences in order to become cost profitable. On the contrary, at $S = 0$ (balanced load scenario), a cooperative cluster of only 8 residences (each provisioned with $(N_{PV} = 1, N_B = 1)$) is required. This variation of annual profit with cluster size when $C_{sh} = C_{buy}$ is shown in Fig. 7.4d. It may be noted that while variation in C_{sh} results in a change in optimal cluster size to become cost profitable, the optimal cluster size to achieve self-sustainability does not change (given in Table 7.1). This is due to the fact that, while higher values of C_{sh} result in lower profits, the energy being shared is still green energy.

Finally through Fig. 7.4e we illustrate the maximum annual energy transferred at a certain skewness level to a residence, attaining self-sustainability. We observe that the quantum of energy transferred increases with increasing skewness and is maximum at $S = 2$ (extreme skewness level). It is also observed that $S = 0$ (balanced load scenario) does not involve energy transfer. This is due to the assumption that the hourly energy harvest at all the residences is assumed equal. The optimum residence cluster size at which a MI-RES system achieves self-sustainability for a given (N_{PV}, N_B) is shown in Table 7.1. From Table 7.1 *we observe that an increase in* N_{PV} *leads to a faster reduction in residence cluster size, as compared to an increase of* N_B. At $(N_{PV}, N_B) = (1, 1)$, we observe the optimal cluster size to be 10. This implies that an optimal MI-RES cluster of ten residences having a joint collocated CAPEX of 35,800 USD $(N_{PV} = 10, N_B = 10)$, corresponds to a per residence CAPEX of 3100 USD $(N_{PV} = 1, N_B = 1)$. This amount is about 13 times lower when compared to a non-energy-cooperative MA-RES system, requiring much higher CAPEX (39,200 USD) to achieve sustainability. Thus achieving a significant reduction in CAPEX per residence, in addition to greenness.

7.6 Summary

The chapter has presented a multi-residence dual-powered, microgrid-based cooperative energy transfer, MI-RES framework, to design self-sustainable next-generation residential frameworks. The designed cooperative mechanism captures and exploits the temporal load inhomogeneities occurring in a dual-powered multi-residential system, thus improving the temporal green energy utilization. The proposed cooperative analytical model has been designed with the green energy storage characterized as a two-state discrete time Markov model. The designed analytical framework has been optimized to compute the optimal quantum of energies to be purchased, sold, or transferred from/to a residence. The proposed cooperative multi-residential MI-RES framework has been compared to a non-cooperative dual-powered, MA-RES framework. The frameworks have been compared in terms of CAPEX incurred per residence towards achieving self-sustainability. The presented cooperative MI-RES framework is expected to pave the way towards self-sustainable green residential societies and motivate existing architectures towards energy cooperation.

Chapter 8
Towards Green AI Aided Future Networks

8.1 Introduction

Networks are prone to many stochastic factors, including traffic dynamics, green energy dynamics, unknown channel state information of the UE-BS links, multi-QoS networks, etc. If not factored in accurately, these stochastic factors result in degrading the system performance. Wireless networks are one of the most lossy (power inefficient) systems, with the wireless channel contributing in reducing the energy efficiency significantly. The wireless environment is also highly prone to dynamic environment, for instance, mobility of the UE, time/frequency selectivity of the channel, blockage effects, and scattering by obstacles.

AI has emerged as a reliable and robust tool to improve the efficiency and robustness of networks, especially with respect to network system design and performance evaluation. In this chapter, we present the key factors inducing stochasticity on network performance, and showcase how AI is being used to improve the system efficiency and accuracy. We also delve as to how AI models can be made energy efficient, and the associated tradeoffs between energy efficiency, computational efficiency, and accuracy of an AI model.

8.2 Green Communication Strategies

State of art green communication strategies can be broadly classified as,

1. BS level
2. Network level
3. Computational level

L.-C. Wang et al., *Green Energy Driven Integrated Smart Grid and Wireless Networks*, Wireless Networks, https://doi.org/10.1007/978-3-032-20509-4_8

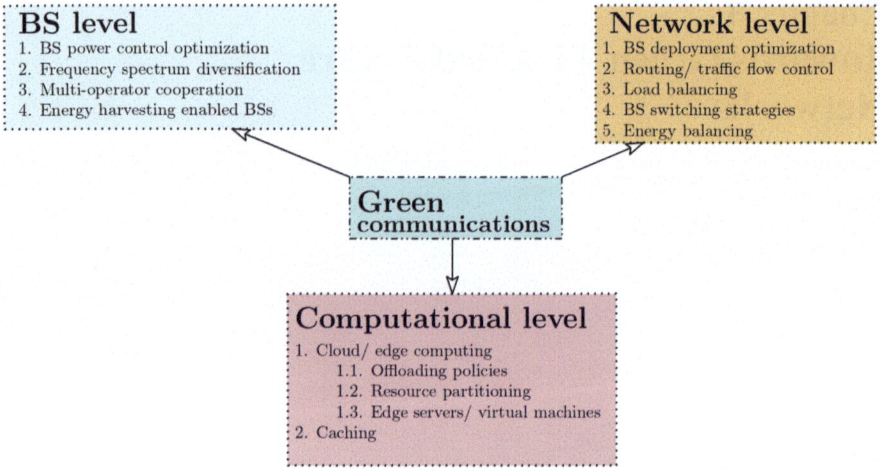

Fig. 8.1 Key research directions towards green communications

Research has been focused from these three broad perspectives towards improving the energy efficiency of the telecommunication system. We present a detailed study of these strategies, as shown in Fig. 8.1 in the upcoming subsections.

8.2.1 BS Level Strategies

The key research directions and challenges associated with BS level strategies is summarized below.

1. BS power control strategies (beamforming, MIMO)
2. More diversified frequency bands
3. Multi operator cooperation
4. Renewable energy powered BSs
5. *Research challenges:* Unknown UE-BS CSI, accurate/ efficient channel estimation, interference mitigation/ avoidance, BS solar provisioning study, UE mobility dynamics

Green communication strategies when implemented at a BS level involve optimizing the physical resources, namely, BS power and frequency allocations [154, 155]. BS power control frameworks involve more sectored and directed beamforming, rather than classical omnidirectional antennas. The frameworks in [154] and [155] presents a reinforcement learning based power control scheme to suppress downlink inter-cell interference and save energy for ultra-dense small cells. MIMO techniques towards increasing the system capacity result in additional

RF-chains, thereby increasing the energy consumption. Hence, optimal resource allocation becomes crucial to avoid resource wastage.

6G aims to design multi-technology and multi-operator compatible RAN. It is aimed to develop efficient spectrum sharing mechanism enabling 4G, 5G, 6G devices to coexist on the same spectrum resource [156]. Such multi-radio access technology coexistence framework can be developed with the aid of AI models in the core network.

Enabling energy harvesting mechanisms at a BS has also developed as a very potential solution towards designing green communication, as has been explained through this monograph. While energy harvest is inherently random, there does exist quasi-periodicity over a small local area. For instance, temporal climatic predictions can be made quite intelligently to guess the wind-speed (in case of wind energy) or solar energy harvest. A statistical analysis of this data over a large time period is expected to be very useful in system performance evaluations (as shown using unsupervised learning model in Chap. 2).

The key challenges in BS level energy optimization lies in accurate estimation the UE-BS CSI link, accurate modeling the LoS/ NLoS of the user considering indoor or outdoor scenarios, capturing the UE mobility dynamics efficiently, and further aiming to mitigate the interference due to multi-BS network at a UE (from a UE perspective).

8.2.2 Network Level Strategies

The key research directions and challenges associated with network level strategies is summarized below.

1. BS deployment optimization
2. Routing/ traffic flow control
3. Load balancing through intelligent UE associations
4. BS switching strategies
5. Energy balancing based cooperative network
6. Multi tier heterogeneous networks (Het-Nets)
7. *Research challenges:* Studying UE/ traffic patterns, studying green energy patterns, User QoS (multi QoS aware routing strategies), effect of mobile 3D BSs on above frameworks (initial access, doppler shift mitigation)

Network level frameworks involve reducing carbon footprint at a network level, through cooperation among the networked nodes. The cooperating among the nodes can be at the physical, data link (medium access control (MAC)), or network layer. Energy efficient network level system designs broadly involve deployment optimization of the network nodes depending on the UE QoS demands, thereby removing over resource provisioning in the network. Such a framework also ensures reduced interference at the UEs considering a scare bandwidth resource available to the operators.

The cellular view of a communication network motivated researchers to design smaller cells, as compared to a classical macro-BS. Such a multi-tier (each tier having a different coverage area) network is termed as a heterogeneous network (Het-Net). Such a Het-Net deployment not only leads to increase in BS deployment, and thereby operator deployment cost, but also increase in network energy. The upcoming 6G aims to deploy aerial three dimensional (3D) BSs like UAV, HAPS, or even satellites over a local coverage area. These 3D BSs are expected to cover a provide ubiquitous connectivity over a significant coverage area, thereby reducing the Het-Net deployments and hence a possible green solution (as also motivated through a HAPS case study in Chap. 6).

The challenge in such aerial BSs is the non-stationarity of the BS itself, resulting in the UE experiencing Doppler shifts. Furthermore, solutions like LEO satellites require a much proper analysis pertaining to handovers, and routing due to the increased path loss. Low altitude UAVs act as good ad hoc BSs, but their viability as permanent ICT infrastructure needs to be carefully studied owing to the devices being energy-constrained.

At the MAC/ link layer, routing or traffic flow control strategies are also useful green communication strategies. It becomes crucial to intelligently route the UE data requests, depending on the UE QoS, intensity of temporal UE load, and the spatio-temporal UE load intensity profile. In the physical layer, load balancing frameworks involve cooperative coverage adjustment (also termed as cell breathing) and BS switching strategies. These strategies involve partial (or complete) shifting the load from a heavily (or lightly loaded) BS to another adjacent BS. These load balancing frameworks assist in homogenizing the load across the networked nodes, thereby improving the resource usage to serve the UEs. A key challenge in load-balancing frameworks is the spatio-temporal randomness of the telecommunication load intensity. Considering the temporal traffic patterns over a long period of time, AI based models can be very useful for accurate prediction of the load intensity at a particular BS. Such look-ahead frameworks are expected to be very useful in proper resource scheduling at the BSs, thereby improving the resource efficiency, in addition to energy efficiency.

Multi-QoS requirements by the UEs when associated with a BS also pose a challenge in optimal resource provisioning by a BS. Cross layer (PHY-MAC) models as in [48] require to be developed, so as to design the scheduler better. AI is expected to assist the BS in estimating the UE CSI, in addition to learning the traffic and UE requirements patterns.

Energy balancing frameworks (as in Chaps. 4 and 5) and joint load-energy balancing frameworks (as in Chap. 5) can be very beneficial especially from a sustainable network design perspective. These strategies are prone to energy dynamics in addition to load dynamics (traffic-energy imbalances discussed in Chap. 3). AI enabled models can be very useful in learning the spatio-temporal energy availability patterns in addition to the load patterns discussed above.

8.2.3 *Computation Level Strategies*

1. Cloud/ edge computing
   (a) offloading policy designs
   (b) computational resource partitioning/ allocation strategy
   (c) Edge based server/ virtual machines
   (d) *Research challenges:* Energy efficiency vs computational efficiency trade-offs, UE QoS aware offloading
2. Content delivery, caching
   (a) Caching policy
   (b) Delivery policy
   (c) *Research challenges:* UE mobility, limited capacity at UE/ edge, unknown CSI

Computation has emerged as a key research area, parallel to communications in recent times owing to the massive amount of computations required to be performed at the BS level/ network core. The BSs generally act as edge servers having limited computational capability, while the core network usually is equipped with a cloud server having a much larger computational capability. Smaller AI models/ computations can be implemented at the edge itself, thereby saving potential delays in communication between the edge and the core (as discussed through lightweight Markov frameworks in Chaps. 2 and 4).

Heavier models are partially implemented at the cloud and partially at the edge. Hence, offloading policy requires to be carefully designed at the edge or at the cloud through proper resource partitioning strategy. From an energy perspective, the cloud servers generally run overtime, consuming massive energy, while the edge nodes consume much lesser energy. Developing edge based servers or virtual machines therefore becomes critical in system design. Caching has emerged as a popular technique to reduce the computational delay. Depending on the user QoS/ QoE, optimal caching policies need to be developed so as to cater to the user QoE in addition to saving resources for computation at the cloud.

In the coming section, we present a detailed overview of the AI models used widely for green communications.

8.3 AI Tools for Green Communication

Classical AI models used for achieving green communications can be classified into the below five categories, shown through Fig. 8.2 [157].

- Heuristic algorithms
- Supervised learning frameworks

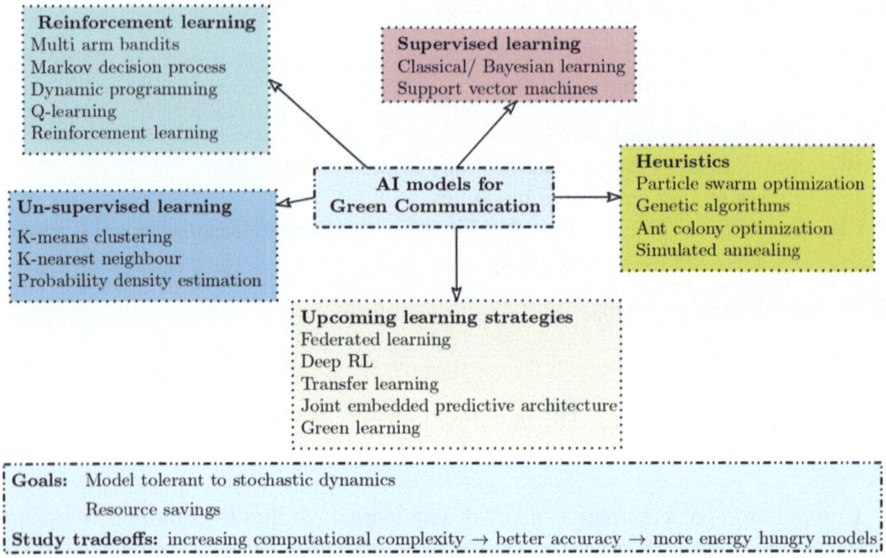

Fig. 8.2 AI models for green communication

- Un-supervised learning frameworks
- Reinforcement learning frameworks
- Deep learning models

In the upcoming subsections, we present a detailed study about these models.

8.3.1 Heuristic AI Models

Heuristic models are iterative algorithms which do not guarantee optimality or convergence. Famous techniques include particle swarm optimizations [158–160], genetic algorithms, and ant colony optimization framework [161]. These frameworks work on NP hard problems and are found to be near optimal in a limited time frame, from a computational perspective. As compared to brute-force greedy search techniques, these heuristics compute much quicker, saving computational resources and time resource at the server.

Optimal edge server/ virtual machine placement has been a very important issue in mobile edge computing (MEC). Energy aware placement problems are in interest so as to reduce the energy consumption. Dynamic consolidation of virtual machines into the minimal number of physical machines is considered as one of the best solutions to manage power consumption. The frameworks in [158–160] have used the particle swarm optimization framework to design and solve energy-aware edge server/ virtual machine placement problems in a MEC setting.

Ant colony optimization is a bio-inspired optimization algorithm that mimics the foraging behavior of ants in nature to solve combinatorial optimization problems, such as the traveling salesman problem or other network routing problems [161]. The framework in [162] presents an ant colony optimization based joint caching and forwarding framework to optimize the power consumption. The adaptive caching framework is a content centric networking model based on the user QoS. The framework in [163] presents a ant colony optimization heuristic to design energy efficient frameworks in a multi-user multi-carrier communication system.

Genetic Algorithms (GAs) are search and optimization techniques inspired by the principles of natural selection and genetics. They belong to the broader class of evolutionary algorithms and are used to find approximate solutions to complex problems where traditional methods struggle [164–166]. The authors in [165] present a container placement problem in container as a service setting, and showcase that energy-saving gain can be achieved through genetic algorithm by optimizing the placement of containers. The authors in [166] present a framework to minimize the number of deployed BSs in a network with coverage constraint, using genetic algorithm thereby improving network energy efficiency.

8.3.2 Classical Supervised Learning Models

Traditional supervised learning frameworks include classical regression analysis [167], classical and Bayesian inference based learning, and support vector machines. The philosophy behind these frameworks lies in training the model with some percentage of data (termed as training data) with the data true labels. Then the remaining data (termed as test data) is then passed on the model, to infer if the trained model is efficiently able to classify the data into its correct label. Classical/ Bayesian inferences also work in the above fashion aiming to minimize the "error probability" through a defined convex cost function.

8.3.3 Un-supervised Learning Models

Un-supervised learning based frameworks like K-means clustering, K-nearest neighbor [168], and probability density estimation techniques do not involve use of true label to train the model. Rather, these model statistically use statistical metrics to cluster data into groups or to fit a given data into summation of Gaussian random variables having usually multiple modes. Such multi-modal analysis is useful when the distribution of the underlying data is very random/ uncertain, requiring a statistical "fit" in addition to minimizing a convex cost function simultaneously (usually MMSE, RMSE minimization).

Recently generative AI models (e.g., autoencoders, diffusion models) have been attractive to the industry. These models also attempt to reconstruct the distribution of the underlying data. Generative modeling includes reconstructing high dimensional inputs, and is very energy intensive. Training costs are usually comparable or even higher than supervised learning models. Joint embedding predictive architectures are expected to be more energy efficient unsupervised learning models, and will be explained in detailed in the next section.

8.3.4 Reinforcement and Deep Learning Models

Reinforcement learning (RL) works with the philosophy of training a model through "positive reinforcements" (rewards) or "negative reinforcements" (penalty). A simple analogy can be 'training a dog' wherein the dog receives rewards/ penalty from the trainer corresponding to its actions during the training process. RL frameworks include multi arm bandits (probabilistically choosing the right arm out of a set of arms), Markov decision process (MDP), dynamic programming, reinforcement learning, and Q-learning.

Deep learning (DL) models are also generally supervised learning models, wherein the data to be trained is passed through a neural network, which works on back propagation and feed forward mechanism [169]. The neural network, comprises of neurons and sigmoidal activation functions per neuron, extracts features at each layer as the data passes through it. Low level features are captured at the lower layers, while high level features are captured at the deeper layers. Examples of deep neural networks are fully connected neural networks and convolutional neural networks. For instance, the authors in [170] presents a deep learning framework to approximate the optimal resource allocation policy to minimize a BS power consumption by optimizing on the bandwidth and transmit power.

DL models are challenging especially from energy efficiency perspective. Deep networks have billions of parameters, some of which are redundant and contribute little to accuracy. These large models require long training hours, resulting in massive energy consumption in data centers. They have poor task generalization, i.e., one model can not be usually used for more than one task. Recent research has focused on combining the RL and DL models together, termed as deep reinforcement learning. We will cover it in detail in the next section.

8.4 Upcoming Green AI Models for 6G

The key goal of green AI is to develop low computation efficiency models (thereby fostering resource savings) with a sufficiently fair accuracy. The model is expected to be tolerant to stochastic network dynamics, and perform robustly even in very random environments. The AI community is currently moving towards five potential frameworks which can realize green AI. These are,

1. Federated learning
2. Deep reinforcement learning
3. Transfer learning
4. Joint embedded predictive architectures
5. Green learning

We explain them in detail in the below subsections.

8.4.1 Federated Learning

Federated learning (FL) is based on the philosophy of collaborative or distributed learning among multiple networked nodes. It is a decentralized algorithm wherein multiple networked devices (termed as clients) collaboratively train a shared model without actually transferring the raw data [171–178]. Instead of communicating the data to a central server, each client locally trains the model and only shares model updates (like gradients or weights) with the central server. Thus, federated learning not only preserves privacy of the local data with each node, but it also reduces the data transfer between the client and server, thereby saving communication as well as computational energy.

The authors in [171] present a FL framework to minimize energy and time consumption for task computation and transmission in a multi UE, HAPS based network. The authors in [172] present a FL framework showcasing latency gain and energy savings in a MEC based setting. The authors in [173] present a dynamic resource optimization framework with imperfect CSI through FL, showcasing energy savings and improved privacy at the UE. To improve the energy performance in IoT frameworks, the authors in [174] present a FL based computation offloading framework through UAV in an IoT scenario. The authors in [175] have presented a FL integrated deep Q-learning framework to solve the partitioning and resource allocation problem in satellite-assisted MEC systems, aiming to minimize the task completion latency. The authors in [177] present a framework to select the most suitable BS to satisfy UE service requests, optimizing QoS and slicing resource utilization through FL. The framework in [178] showcases bandwidth savings in multi UE systems through FL.

The challenge in designing FL based networks is the requirement for statistical heterogeneity, i.e., each client should have a non identical and independent (IID) data distributions. This may also lead to slower convergence and reduced global model performance. Furthermore FL is communication heavy, resulting in possible communication bottlenecks at the server.

8.4.2 Deep Reinforcement Learning

Deep RL (DRL) is based on the philosophy that a networked agent learns how to act in an environment by trial and error, using a deep neural network to guide its decisions. It is a combination of classical RL and deep neural networks. Deep RL models are expected to be very tolerant and robust especially to fast changing dynamics. Its key advantages are,

- Deep RL is based on policy optimization and can naturally learn stochastic policies.
- Hence, it is robust in decision making even under stochastic environments.
- As combined with deep neural network, the RL framework can be trained to diverse, noisy environments
- It encourages exploration, as compared to overfitting which is quite prevalent in certain models.

The authors in [179] present a deep RL framework to maximize energy efficiency in a UAV network, with joint consideration for communications coverage, fairness, energy consumption and connectivity. The authors in [180] study the joint problem of dynamic multi-UAV altitude control and multi-cell wireless channel access management of IoT devices. The paper showcases that the IoT devices are dynamically controlled to attain the maximal long-term network capacity while ensuring energy sustainability of UAVs. The authors in [180] present a DRL based framework to design a task scheduling policy that minimizes offloading and computing delay of all tasks given the UAV energy capacity constraint.

The authors in [181] present a DRL framework to optimize the grid energy and traffic drop rate through coordination among the networked BSs. They showcase that the distributed DRL solution provides higher network performance, better adaptation to the changing environment, and higher cost savings. The authors in [182] utilize DRL framework to allocate resources to users in a near optimal way in a NOMA setting. The authors in [183] present a DRL framework to optimize cell range expansion bias and transmit power allocation, as well as appropriate frequency bands for multi-flow transmissions to a UE in a multi-tier Het-Net. The authors in [154, 184] propose a DRL based power control scheme to suppress downlink inter-cell interference and save energy for ultra-dense small cells. The authors in [185] present a DRL framework to balance the energy consumption in local computation and data offloading in a MEC setting. To maximize the energy efficiency in MEC offloading, the DRL framework is customized to learn the optimal transmission scheduling and workload allocation.

The challenges in implementing DRL models include, lack of exhaustive dataset, the tradeoffs between exploitation and exploration in RL, convergence issues, and partial observability in many environments.

8.4.3 Transfer Learning

Transfer Learning is a machine learning technique wherein a model trained on one task is reused or to another but related task. The concept lies in aversion to train a model from scratch for all tasks [186]. As a real-life analogy, assume someone "learns" to play tennis, involving key aspects like, hand-eye coordination, swing timing, or footwork. The same aspects can be used to learn badminton (another racquet game). Hence, a person who knows to play tennis, can quickly learn badminton or other raquet sports, and not need to learn from scratch.

Unlike classical AI models which need to be trained from zero, transfer learning is based on fine tuning the new model on existing training model available. Hence, it can significantly reduce the computation resources and required training data (thus energy gains), resulting in extended and accelerated learning for the new application. In wireless communications, recent frameworks in [170, 187–189] have applied transfer learning in wireless applications.

The authors in [187, 188] have presented a framework for dynamic sectorization in BS, thereby enhancing the network energy efficiency. The framework exploits previously estimated traffic statistics through transfer learning to automate the sectorization process by a BS, not only speeding up the training process, but also improving the energy efficiency. The transfer learning framework is extended to switching BS ON/OFF based on the transfer learning aided traffic statistics learnt by the core network. The authors in [170] present a framework to optimize the radio resource allocation in a BS, towards minimizing the power consumption. Transfer learning has been used to update the neural networks for non-stationary wireless channels. The framework in [189] presents a computational offloading problem of IoT networks to cloud RAN. Transfer learning is used here to fine tune the neural network in dynamic IoT systems.

The challenges associated with transfer learning are that the tasks to be learnt should not be very dis-similar, or show very less correlation. Performance degradation is very likely in such scenarios. Proper fine tuning is also very crucial, else may lead to overfitting.

8.4.4 Joint Embedding Predictive Architecture

Joint Embedding Predictive Architecture (JEPA) is a framework proposed to advance self-supervised learning, especially as an alternative to traditional supervised or generative methods for building general purpose AI systems. The core philosophy is to develop autonomous intelligence without heavy reliance on labeled data or pure generative models [190, 191].

JEPA model is energy-efficient primarily because it avoids generating or reconstructing raw high-dimensional data (as in generative models like auto encoders), focusing instead on predicting compact, abstract representations in latent space.

This significantly reduces computational cost. The key points summarizing energy efficient designs in JEPA based modeling is detailed below.

- Traditional generative models (e.g., autoencoders, diffusion models, transformers) try to generate data at the pixel level. Hence, traditional generative models are very fine grained, high dimensional, and expensive to compute. In contrast, JEPA models predict low dimensional vectors, which are smaller, more abstract, and cheaper to compute.
- There are no need for decoders during inference in JEPA models. Once trained, a JEPA model can discard the target encoder and predictor, keeping only the encoder for downstream tasks (e.g., classification, reinforcement learning). This modularity makes inference cheaper than traditional generative models like transformers which require full context windows and attention heads.
- Training JEPA models can lead to faster convergence, require less data augmentation, and avoid fine grained predictions.

The challenges associated with JEPA models are briefly discussed. JEPA operates entirely in latent space, predicting abstract embeddings rather than raw data. The accuracy of learning is very dependent on how well this latent space is structured. Further, JEPA relies on predicting future/context embeddings, which are themselves learned and potentially unstable. Also, without careful architectural or training design (e.g., stop-gradient, normalization), JEPA models are prone to learning trivial mappings.

Therefore, JEPA is a very exciting upcoming energy-efficient and abstraction-focused architecture. It requires careful training strategies, stronger empirical validation, and mechanisms to stabilize and interpret the latent space.

8.4.5 Green Learning

Recently, green learning has been proposed as an energy efficient AI model by the authors in [192]. The green learning models are characterized by low carbon footprints, lightweight modelings, low computational complexity, and logical transparency. The proposed model combines unsupervised and supervised learning intelligently in a cascaded fashion, for feature extraction and decision making.

The green learning framework in [192] goes against traditional neural network design, and does not have a computational neuron as the base unit. The system consists of three cascaded learning modules, explained below.

- *Unsupervised representation learning*: This module is used to for dimensionality reduction. The tools include discrete cosine transform and principal component analysis.
- *Supervised feature learning*: This module is used to select useful features from a large set of representations.

- *Supervised decision learning*: This module is used to conduct feature ensembles and the final classification task.

The challenges in green learning based models include robustness and vulnerability to cyber attacks, need for a more comprehensive risk assessment, ability to handle multiple domain of data, and integration with human cognition.

8.5 Summary

Developing computationally light and efficient AI models is still an ongoing research. Energy consumption by different AI models needs to be properly studied and considered to design lightweight AI models. The key question we need to ask is if we can decrease the training data and algorithmic complexity, without compromising on the accuracy. Therefore, the tradeoff between energy efficiency and network performance remains the core theme. Cyber security has emerged as a key component in future network designing. Security together with sustainability might eventually be the key factor deciding whether AI enabled 6G communications can be sustained for a long term among the common masses.

Chapter 9
Conclusion

This monograph was aimed at demonstrating the feasibility of designing 'sustainable' sixth generation future networks through dynamic network cooperation. The term 'sustainability' has been coined to include both energy as well as economic sustainability. The research presented through the monograph has showcased the integration of renewable energy with smart grid connected networks. With the upcoming 6G networks expected to be centered on providing ubiquitous coverage, sustainability is expected to be a key constraint, influencing the ubiquitous deployability of 6G systems.

The core idea in the monograph has been to enable each grid user as a potential energy source to the power grid, thereby revising the conventional idea of power grids. We have not only looked at energy-prosumer designs, but also aimed at energy-producer designs, wherein the distributed system of grid users, act as energy-producer to the power grid.

The monograph has motivated the need sustainability in the next-generation communications, owing to expected increase in power consumption in 5G and beyond networks, inflexibility of traditional power grid, and non-scalability of stand-alone green communication system. The monograph has also showcased futuristic aerial-terrestrial network design and the flexibility of extending the generic load-energy balancing concept to general consumer utility networks. Futuristic green AI aided network design is also shown exploring the tradeoff between energy efficiency and computational complexity. Integrating the smart grid with renewable energy sources, is expected to assist in scaling the green solutions by opening new avenues of operator revenue.

Ubiquitous connectivity, though pertinent, gets limited due to sustainability issues. We believe that security together with sustainability shall be a key deciding factor as to whether 6G communications can be sustained in the long run among the common masses.

L.-C. Wang et al., *Green Energy Driven Integrated Smart Grid and Wireless Networks*, Wireless Networks, https://doi.org/10.1007/978-3-032-20509-4_9

Appendix A
Proof of Optimal Grid Energy Consumption

Proof To solve problem $\mathcal{P}2$, the Lagrangian of the primal problem $\mathcal{P}2$ is defined as follows:

$$\begin{aligned}\mathbb{L}(\mathcal{S}_b, \lambda, \mu, \gamma) &= \sum_{b=1}^{ED} (\mathcal{S}_b(t))^2 + \lambda \left(\sum_{b=1}^{ED} (\mathcal{S}_b(t) - \mathcal{D}_b(t)) \right) \\ &+ \mu \left(\sum_{b=1}^{ED} \mathcal{S}_b(t) - \min \left\{ \sum_{b'=1}^{ES} \mathcal{E}_{b'}^{S}(t), \sum_{b=1}^{ED} \mathcal{D}_b(t) \right\} \right) - \gamma \left(\sum_{b=1}^{ED} \mathcal{S}_b(t) \right).\end{aligned} \tag{A.1}$$

Solving for $\partial \mathbb{L}(\mathcal{S}_b, \lambda, \mu, \gamma)/\partial \mathcal{S}_b = 0$, the following expression is obtained.

$$\mathcal{S}_b^*(t) = \frac{(\gamma - \lambda - \mu)}{2}. \tag{A.2}$$

$\mathcal{S}_b^*(t)$ represents the primal solution of problem $\mathcal{P}2$. To solve for the Lagrange multipliers, the value of $\mathcal{S}_b^*(t)$ in $\mathbb{L}(\mathcal{S}_b, \lambda, \mu, \gamma)$ is substituted, to obtain the dual function $g(\lambda, \mu, \gamma)$. The dual function obtained after substituting the primal solution in the Lagrangian is given below.

L.-C. Wang et al., *Green Energy Driven Integrated Smart Grid and Wireless Networks*, Wireless Networks, https://doi.org/10.1007/978-3-032-20509-4

$$
\begin{aligned}
g(\gamma, \lambda, \mu) = & \frac{(\gamma - \lambda - \mu)^2 ED}{4} + \frac{\lambda(\gamma - \lambda - \mu) ED}{2} - \lambda \sum_{b=1}^{ED} \mathcal{D}_b(t) \\
& + \frac{\mu(\gamma - \lambda - \mu) ED}{2} - \mu \min\left\{ \sum_{b'=1}^{ES} \mathcal{E}_{b'}^S(t), \sum_{b=1}^{ED} \mathcal{D}_b(t) \right\} \\
& - \frac{\gamma(\gamma - \lambda - \mu) ED}{2} \\
= & \frac{-(\gamma - \lambda - \mu)^2 ED}{4} - \lambda \sum_{b=1}^{ED} \mathcal{D}_b(t) \\
& - \mu \min\left\{ \sum_{b'=1}^{ES} \mathcal{E}_{b'}^S(t), \sum_{b=1}^{ED} \mathcal{D}_b(t) \right\}.
\end{aligned} \tag{A.3}
$$

Solving the dual function for λ, i.e., $\partial g(\gamma, \lambda, \mu)/\partial \lambda = 0$, the following expression is obtained.

$$
(\gamma - \lambda - \mu)^* = \frac{2 \sum_{b=1}^{ED} \mathcal{D}_b(t)}{ED}. \tag{A.4}
$$

Substituting the derived dual solution, $(\gamma - \lambda - \mu)^*$, in the primal solution (A.2) gives

$$
\mathcal{S}_b^*(t) = \sum_{b=1}^{ED} \mathcal{D}_b(t)/ED. \tag{A.5}
$$

Now, solving the dual function for γ, i.e., $\partial g(\gamma, \lambda, \mu)/\partial \gamma = 0$, gives

$$
(\gamma - \lambda - \mu)^* = 0. \tag{A.6}
$$

On substituting (A.6) in (A.2), it gives $\mathcal{S}_b^*(t) = 0$.

Finally, solving the dual function for μ, i.e., $\partial g(\gamma, \lambda, \mu)/\partial \mu = 0$, the following expression is obtained,

$$
(\gamma - \lambda - \mu) = \frac{2 \min\left\{ \sum_{b'=1}^{ES} \mathcal{E}_{b'}^S(t), \sum_{b=1}^{ED} \mathcal{D}_b(t) \right\}}{ED}, \tag{A.7}
$$

which on substituting in (A.2) gives

$$
\mathcal{S}_b^*(t) = \frac{\min\left\{ \sum_{b'=1}^{ES} \mathcal{E}_{b'}^S(t), \sum_{b=1}^{ED} \mathcal{D}_b(t) \right\}}{ED}. \tag{A.8}
$$

It is observed that (A.5) $\subset$ (A.8). Hence (A.8) is the optimum solution of the primal problem, $\mathcal{P}2$. (A.8) can also be expressed as

$$\mathcal{S}_b^*(t) = \begin{cases} \sum_{b=1}^{ED} \mathcal{D}_b(t)/ED, \text{ if } \sum_{b=1}^{ED} \mathcal{D}_b(t) \leq \sum_{b'=1}^{ES} \mathcal{E}_{b'}^S(t) \\ \sum_{b'=1}^{ES} \mathcal{E}_{b'}^S(t)/ED, \text{ if } \sum_{b=1}^{ED} \mathcal{D}_b(t) > \sum_{b'=1}^{ES} \mathcal{E}_{b'}^S(t) \end{cases} \tag{A.9}$$

From (A.9) it is observed that $\mathcal{S}_b^*(t) = \sum_{b=1}^{ED} \mathcal{D}_b(t)/ED$ if $\sum_{b=1}^{ED} \mathcal{D}_b(t) \leq \sum_{b'=1}^{ES} \mathcal{E}_{b'}^S(t)$. Intuitively, since the aim is to reduce the grid-energy procurement, $\mathcal{S}_b^*(t) = \mathcal{D}_b(t)$ if $\sum_{b=1}^{ED} \mathcal{D}_b(t) \leq \sum_{b'=1}^{ES} \mathcal{E}_{b'}^S(t)$. Thus, it is observed that $\mathcal{S}_b(t) = \mathcal{D}_b(t) = \sum_{b=1}^{ED} \mathcal{D}_b(t)/ED$. Hence, for $\sum_{b=1}^{ED} \mathcal{D}_b(t) > \sum_{b'=1}^{ES} \mathcal{E}_{b'}^S(t)$, $\mathcal{S}_b(t)$ can be expressed as

$$\mathcal{S}_b(t) = \frac{\mathcal{D}_b(t) \sum_{b'=1}^{ES} \mathcal{E}_{b'}^S(t)}{\sum_{b=1}^{ED} \mathcal{D}_b(t)}. \tag{A.10}$$

Thus, (A.9) can be rewritten as

$$\mathcal{S}_b^*(t) = \begin{cases} \mathcal{D}_b(t), \text{ if } \sum_{b=1}^{ED} \mathcal{D}_b(t) \leq \sum_{b'=1}^{ES} \mathcal{E}_{b'}^S(t) \\ \mathcal{D}_b(t) \sum_{b'=1}^{ES} \mathcal{E}_{b'}^S(t) / \sum_{b=1}^{ED} \mathcal{D}_b(t), \\ \qquad \text{if } \sum_{b=1}^{ED} \mathcal{D}_b(t) > \sum_{b'=1}^{ES} \mathcal{E}_{b'}^S(t) \end{cases} \tag{A.11}$$

or,

$$\mathcal{S}_b^*(t) = \frac{\mathcal{D}_b(t) \min\left\{\sum_{b'=1}^{ES} \mathcal{E}_{b'}^S(t), \sum_{b=1}^{ED} \mathcal{D}_b(t)\right\}}{\sum_{b=1}^{ED} \mathcal{D}_b(t)}. \tag{A.12}$$

The above equation (A.12) represents the hourly optimal value of energy to be shared from energy-sufficient BSs to a deficient BS. Since $\mathcal{G}_b(t) = \mathcal{D}_b(t) - \mathcal{S}_b(t)$, from (A.12) the following expression is obtained,

$$\mathcal{G}_b^*(t) = \begin{cases} 0, \text{ if } \sum_{b=1}^{ED} \mathcal{D}_b(t) \leq \sum_{b'=1}^{ES} \mathcal{E}_{b'}^S(t) \\ \mathcal{D}_b(t) \left(\frac{\sum_{b=1}^{ED} \mathcal{D}_b(t) - \sum_{b'=1}^{ES} \mathcal{E}_{b'}^S(t)}{\sum_{b=1}^{ED} \mathcal{D}_b(t)} \right), \\ \qquad \text{if } \sum_{b'=1}^{ES} \mathcal{E}_{b'}^S(t) < \sum_{b=1}^{ED} \mathcal{D}_b(t). \end{cases} \tag{A.13}$$

This completes the proof. □

Appendix B
Proof of Consecutive Energy Outage

Proof The probability that the system will transit to the deficit state (D) from any state in the next time slot is given as

$$P_1 = \pi_D \mathbb{P}_{\overline{D}D} \mathbb{P}_{D\overline{D}}. \tag{B.1}$$

Similarly, the probability that the system will transit to state D in the next two and three consecutive time slots, respectively is

$$\begin{aligned} P_2 &= \pi_D \mathbb{P}_{\overline{D}D} \mathbb{P}_{DD} \mathbb{P}_{D\overline{D}} \\ P_3 &= \pi_D \mathbb{P}_{\overline{D}D} \mathbb{P}^2_{DD} \mathbb{P}_{D\overline{D}} \end{aligned} \tag{B.2}$$

Thus, the probability that the EC system will transit to state D in the next k consecutive time slots is

$$P_k = \pi_D \mathbb{P}_{\overline{D}D} \mathbb{P}^{k-1}_{DD} \mathbb{P}_{D\overline{D}} \tag{B.3}$$

This completes the proof. □

L.-C. Wang et al., *Green Energy Driven Integrated Smart Grid and Wireless Networks*, Wireless Networks, https://doi.org/10.1007/978-3-032-20509-4

Appendix C
Proof of Optimal Grid Procurement in Operator Revenue Maximization

Proof The Lagrangian corresponding to the above primal problem is shown below.

$$\begin{aligned}\mathbb{L}\left(\mathcal{S}_b(t), \gamma, \mu\right) =&(C_b - C_{se} - C_{sh})^2 \sum_{b=1}^{ED} (\mathcal{S}_b)^2 + \kappa - \gamma \sum_{b=1}^{ED} \mathcal{S}_b \\ &+ \mu \left(\sum_{b=1}^{ED} \mathcal{S}_b - \min \left\{ \sum_{b'=1}^{ES} \mathcal{E}_{b'}^S, \sum_{b=1}^{ED} \mathcal{D}_b \right\} \right).\end{aligned} \tag{C.1}$$

Solving for $\partial \mathbb{L}\left(\mathcal{S}_b(t), \gamma, \mu\right) / \partial \mathcal{S}_b = 0$ yields

$$\mathcal{S}_b^* = \frac{(\gamma - \mu)}{2 \times (C_b - C_{se} - C_{sh})^2}. \tag{C.2}$$

$\mathcal{S}_b^*$ represents the primal solution of $\mathcal{P}5$. The primal solution in (C.2) is used to obtain the dual, which is given as

$$\begin{aligned}g(\gamma, \mu) =&\frac{ED(\gamma - \mu)^2}{4(C_b - C_{se} - C_{sh})^2} + \kappa + \frac{(\mu\gamma - \mu^2)ED}{2(C_b - C_{se} - C_{sh})^2} \\ &+ \frac{(\mu\gamma - \gamma^2)ED}{2(C_b - C_{se} - C_{sh})^2} - \mu \min \left\{ \sum_{b'=1}^{ES} \mathcal{E}_{b'}^S, \sum_{b=1}^{ED} \mathcal{D}_b \right\}.\end{aligned} \tag{C.3}$$

Solving for $\partial g(\gamma, \mu)/\partial \gamma = 0$ gives,

$$\frac{-ED(\gamma - \mu)}{2(C_b - C_{se} - C_{sh})^2} = 0. \tag{C.4}$$

L.-C. Wang et al., *Green Energy Driven Integrated Smart Grid and Wireless Networks*, Wireless Networks, https://doi.org/10.1007/978-3-032-20509-4

This results in $(\gamma - \mu) = 0$, or further $\mathcal{S}_b^* = 0$. Solving for $\partial g(\gamma, \mu)/\partial \mu = 0$, gives,

$$\frac{(\gamma - \mu)ED}{2(C_b - C_{se} - C_{sh})^2} - \min\left\{\sum_{b'=1}^{ES} \mathcal{E}_{b'}^S, \sum_{b=1}^{ED} \mathcal{D}_b\right\} = 0$$
$$\text{or,}\ \ (\gamma - \mu) = \frac{\min\left\{\sum_{b'=1}^{ES} \mathcal{E}_{b'}^S, \sum_{b=1}^{ED} \mathcal{D}_b\right\} \times 2(C_b - C_{se} - C_{sh})^2}{ED} \tag{C.5}$$

The expression (C.5) represents the dual solution of $\mathcal{P}5$. Substituting (C.5) in the primal solution (C.2) gives,

$$\mathcal{S}_b^* = \frac{\min\left\{\sum_{b'=1}^{ES} \mathcal{E}_{b'}^S(t), \sum_{b=1}^{ED} \mathcal{D}_b(t)\right\}}{ED}$$
$$\text{or, } \mathcal{S}_b^*(t) = \begin{cases} \mathcal{D}_b(t), \text{ if } \sum_{b=1}^{ED} \mathcal{D}_b(t) \leq \sum_{b'=1}^{ES} \mathcal{E}_{b'}^S(t) \\ \left(\mathcal{D}_b(t) \sum_{b'=1}^{ES} \mathcal{E}_{b'}^S(t)\right) / \sum_{b=1}^{ED} \mathcal{D}_b(t), \\ \qquad \text{if } \sum_{b'=1}^{ES} \mathcal{E}_{b'}^S(t) < \sum_{b=1}^{ED} \mathcal{D}_b(t). \end{cases} \tag{C.6}$$

From (4.10) and (C.6), it can be observed that the optimal energy shared (and the corresponding grid energy procurement) to an energy-deficient BS in grid energy procurement minimization strategy is the same as that in operator revenue maximization strategy. Hence it is proven that for a given CAPEX, the OPEX incurred in both optimization problems are identical.

This completes the proof. □

References

1. V. Chamola, B. Sikdar, B. Krishnamachari, IEEE Trans. Commun. **65**(3), 1092 (2017)
2. A. Balakrishnan, S. De, L.C. Wang, IEEE Trans. Commun. **70**(10), 6977 (2022)
3. P. Yang, Y. Xiao, M. Xiao, S. Li, IEEE Netw. **33**(4), 70 (2019)
4. International Telecommunication Union (ITU),Framework and overall objectives of the future development of IMT for 2030 and beyond. Recommendation, ITU-R M.2160-0 (11/2023). https://www.itu.int/en/ITU-T/publications/Pages/recs.aspx
5. Nokia: Spectrum for 6G explained. https://www.nokia.com/6g/spectrum-for-6G-explained/
6. Z. Li, M.A. Uusitalo, H. Shariatmadari, B. Singh, *Proc. IEEE ISWCS* (2018), pp. 1–6
7. S.R. Pokhrel, J. Ding, J. Park, O.S. Park, J. Choi, IEEE Access **8**, 131796 (2020)
8. A. Anand, G. de Veciana, S. Shakkottai, IEEE/ACM Trans. Netw. **28**(2), 477 (2020)
9. P. Popovski, K.F. Trillingsgaard, O. Simeone, G. Durisi, IEEE Access **6**, 55765 (2018)
10. O.O. Erunkulu, A.M. Zungeru, C.K. Lebekwe, M. Mosalaosi, J.M. Chuma, IEEE Access **9**, 97251 (2021)
11. H. Ullah, N. Gopalakrishnan Nair, A. Moore, C. Nugent, P. Muschamp, M. Cuevas, IEEE Access **7**, 37251 (2019)
12. M. Boban, A. Kousaridas, K. Manolakis, J. Eichinger, W. Xu, CoRR abs/1712.01754 (2017). http://arxiv.org/abs/1712.01754
13. D. Zhang, J.J.P.C. Rodrigues, Y. Zhai, T. Sato, IEEE Commun. Mag. **59**(9), 80 (2021)
14. A. Mahmood, L. Beltramelli, S. Fakhrul Abedin, S. Zeb, N.I. Mowla, S.A. Hassan, E. Sisinni, M. Gidlund, IEEE Trans. Ind. Inform. **18**(6), 4122 (2022)
15. W. Mao, Z. Zhao, Z. Chang, G. Min, W. Gao, IEEE Trans. Ind. Inform. **17**(11), 7225 (2021)
16. GSMA report: The Mobile Economy 2025. https://www.gsma.com/solutions-and-impact/connectivity-for-good/mobile-economy/wp-content/uploads/2025/04/030325-The-Mobile-Economy-2025.pdf
17. A. Gaur, A. Balakrishnan, S. Srirangarajan, S. De, P.H. Tseng, K.T. Feng, IEEE Wirel. Commun. Lett. **13**(12), 3390 (2024)
18. K.B. Letaief, W. Chen, Y. Shi, J. Zhang, Y.J.A. Zhang, IEEE Commun. Mag. **57**(8), 84 (2019)
19. J. Hoydis, F.A. Aoudia, A. Valcarce, H. Viswanathan, IEEE Commun. Mag. **59**(5), 76 (2021)
20. B. Mao, F. Tang, Y. Kawamoto, N. Kato, IEEE Commun. Surv. Tutor. **24**(1), 210 (2022)
21. Z. Zhang, L. Dai, X. Chen, C. Liu, F. Yang, R. Schober, H.V. Poor, IEEE Trans. Commun. **71**(3), 1707 (2023)
22. S. Basharat, S.A. Hassan, H. Pervaiz, A. Mahmood, Z. Ding, M. Gidlund, IEEE Wirel. Commun. **28**(6), 184 (2021)
23. C. Pan, H. Ren, K. Wang, J.F. Kolb, M. Elkashlan, M. Chen, M. Di Renzo, Y. Hao, J. Wang, A.L. Swindlehurst, X. You, L. Hanzo, IEEE Commun. Mag. **59**(6), 14 (2021)

L.-C. Wang et al., *Green Energy Driven Integrated Smart Grid and Wireless Networks*, Wireless Networks, https://doi.org/10.1007/978-3-032-20509-4

24. Y. Wang, W. Zhang, Y. Chen, C.X. Wang, J. Sun, IEEE Commun. Lett. **26**(6), 1413 (2022)
25. M. Giordani, M. Zorzi, IEEE Netw. **35**(2), 244 (2021)
26. Z. Xiao, Z. Han, A. Nallanathan, O.A. Dobre, B. Clerckx, J. Choi, C. He, W. Tong, IEEE J. Sel. Areas Commun. **40**(10), 2773 (2022)
27. N.N. Dao, Q.V. Pham, N.H. Tu, T.T. Thanh, V.N.Q. Bao, D.S. Lakew, S. Cho, IEEE Commun. Surv. Tutor. **23**(2), 1193 (2021)
28. W. Lee, IEEE Trans. Veh. Technol. **39**(3), 187 (1990)
29. J.G. Andrews, S. Buzzi, W. Choi, S.V. Hanly, A. Lozano, A.C.K. Soong, J.C. Zhang, IEEE J. Sel. Areas Commun. **32**(6), 1065 (2014)
30. J.G. Andrews, X. Zhang, G.D. Durgin, A.K. Gupta, IEEE Commun. Mag. **54**(10), 184 (2016)
31. W. Feng, Y. Wang, D. Lin, N. Ge, J. Lu, S. Li, IEEE J. Sel. Areas Commun. **35**(7), 1459 (2017)
32. G. Yang, M. Xiao, H.V. Poor, IEEE Trans. Commun. **66**(8), 3526 (2018)
33. VIAVI Solutions: What is 5G Energy Consumption? https://www.viavisolutions.com/en-us/resources/learning-center/what-5g-energy-consumption
34. C. Han, T. Harrold, S. Armour, I. Krikidis, S. Videv, P.M. Grant, H. Haas, J.S. Thompson, I. Ku, C.X. Wang, T.A. Le, M.R. Nakhai, J. Zhang, L. Hanzo, IEEE Commun. Mag. **49**(6), 46 (2011)
35. Ericsson: On the road to breaking the energy curve. https://www.ericsson.com/en/about-us/sustainability-and-corporate-responsibility/environment/product-energy-performance
36. Telecoms industry sets pathway to achieve net zero emissions. https://renewablesnow.com/news/telecoms-industry-sets-pathway-to-achieve-net-zero-emissions-688962/
37. M. Kim, T. Klein, E. Soljanin, J. Barros, M. Médard, Mob. Netw. Appl. **19**, 790 (2014)
38. M. Kim, T. Klein, E. Soljanin, J. Barros, M. Médard, *Proc. MONAMI*, Cork, September (Springer, Berlin, 2013), pp. 1–14
39. Reuters: US power grid to rely on renewable sources, EIA says, April 14, 2023. https://www.reuters.com/business/energy/us-power-grid-rely-renewable-sources-eia-says-2023-04-13/
40. T. Han, N. Ansari, IEEE Wirel. Commun. **21**(1), 90 (2014)
41. M.A. Ahmed, A.M. Eltamaly, M.A. Alotaibi, A.I. Alolah, Y.C. Kim, IEEE Access **8**, 40180 (2020)
42. F.R. Yu, P. Zhang, W. Xiao, P. Choudhury, IEEE Netw. **25**(5), 22 (2011)
43. S.K. Viswanath, C. Yuen, W. Tushar, W.T. Li, C.K. Wen, K. Hu, C. Chen, X. Liu, IEEE Wirel. Commun. **23**(5), 90 (2016)
44. V. Chamola, B. Sikdar, IEEE Commun. Mag. **54**(5), 108 (2016)
45. V. Chamola, B. Sikdar, *Proc. IEEE ICC* (2015), pp. 172–177
46. J. Xu, L. Duan, R. Zhang, IEEE Commun. Mag. **53**(5), 257 (2015)
47. U. Boeke, M. Wendt, *2015 IEEE First International Conference on DC Microgrids (ICDCM)* (2015), pp. 210–214
48. A. Balakrishnan, S. De, L.C. Wang, IEEE Trans. Green Commun. Netw. **5**(4), 1791 (2021)
49. A. Balakrishnan, S. De, L.C. Wang, *Proc. IEEE GLOBECOM* (2020), pp. 1–6
50. A. Balakrishnan, S. De, L.C. Wang, *Proc. IEEE GLOBECOM* (2022), pp. 1–6
51. E. Oh, K. Son, B. Krishnamachari, IEEE Trans. Wirel. Commun. **12**(5), 2126 (2013)
52. D. Renga, H. Al Haj Hassan, M. Meo, L. Nuaymi, IEEE Trans. Green Commun. Netw. **2**(3), 868 (2018)
53. Z. Niu, Y. Wu, J. Gong, Z. Yang, IEEE Commun. Mag. **48**(11), 74 (2010)
54. H. Ghazzai, M.J. Farooq, A. Alsharoa, E. Yaacoub, A. Kadri, M.S. Alouini, IEEE Trans. Veh. Technol. **66**(7), 5879 (2016)
55. J. Gong, J.S. Thompson, S. Zhou, Z. Niu, IEEE Trans. Commun. **62**(11), 3801 (2014)
56. A.Y. Awan, M. Ali, M. Naeem, F. Qamar, M.N. Sial, IEEE Trans. Ind. Inform. **16**(3), 1914 (2020)
57. P. Yu, M. Yang, A. Xiong, Y. Ding, W. Li, X. Qiu, L. Meng, M. Kadoch, M. Cheriet, IEEE Trans. Ind. Inform. **18**(1), 520 (2022)
58. N. Saxena, A. Roy, H. Kim, IEEE Trans. Ind. Inform. **13**(3), 1471 (2017)
59. T. Han, N. Ansari, IEEE Trans. Wirel. Commun. **12**(8), 3872 (2013)

60. T. Zhou, Y. Yue, D. Qin, X. Nie, X. Li, C. Li, IEEE Internet Things J. **9**(19), 18695 (2022)
61. H. Jo, Y.J. Sang, P. Xia, J.G. Andrews, IEEE Trans. Wirel. Commun. **11**(10), 3484 (2012)
62. H. Wu, C. Qiao, S. De, O. Tonguz, IEEE J. Sel. Areas Commun. **19**(10), 2105 (2001)
63. A. Goldsmith, S. Wicker, IEEE Wirel. Commun. **9**(4), 8 (2002)
64. S. Hu, X. Chen, W. Ni, X. Wang, E. Hossain, IEEE Trans. Green Commun. Netw. **4**, 461 (2020)
65. J. Wang, C. Jiang, Z. Han, Y. Ren, L. Hanzo, IEEE J. Sel. Areas Commun. **34**(12), 3785 (2016)
66. H. Al Haj Hassan, D. Renga, M. Meo, L. Nuaymi, IEEE Trans. Green Commun. Netw. **3**, 381 (2019)
67. S. Suman, S. De, IEEE Trans. Sustain. Comput. **5**(3), 438 (2020)
68. S. Suman, S. De, *Proc. IEEE WoWMoM*, Macau (June 2017), pp. 1–8
69. V. Chamola, B. Krishnamachari, B. Sikdar, IEEE Syst. J. **12**(3), 2622 (2018)
70. H. Al Haj , K. Deep Singh, L. Nuaymi, *Proc. IEEE OnlineGreenComm*, Piscataway (Nov. 2016), pp. 1–6
71. L. Chen, B. Wang, X. Chen, X. Zhang, D. Yang, *Proc IEEE INFOCOM Wkshps.*, Shanghai (Apr. 2011), pp. 91–96
72. L. Xiang, X. Ge, C.X. Wang, F.Y. Li, F. Reichert, IEEE Trans. Wirel. Commun. **12**(3), 961 (2013)
73. X. Chen, J. Wu, Y. Cai, H. Zhang, T. Chen, IEEE J. Sel. Areas Commun. **33**(4), 627 (2015)
74. D. Lee, S. Zhou, X. Zhong, Z. Niu, X. Zhou, H. Zhang, IEEE Wirel. Commun. **21**(1), 80 (2014)
75. J. Wu, S. Zhou, Z. Niu, IEEE Trans. Wirel. Commun. **12**(8), 4196 (2013)
76. Y. Chia, S. Sun, R. Zhang, IEEE Trans. Wirel. Commun. **13**(12), 6996 (2014)
77. M.J. Farooq, H. Ghazzai, A. Kadri, H. ElSawy, M.S. Alouini, IEEE Trans. Commun. **65**(2), 918 (2017)
78. A. Balakrishnan, S. De, L.C. Wang, *Proc. IEEE GLOBECOM* (2021), pp. 1–6
79. H.S. Lee, J.W. Lee, IEEE Syst. J. **14**(1), 132 (2020)
80. G. Auer, V. Giannini, C. Desset, I. Godor, P. Skillermark, M. Olsson, M.A. Imran, D. Sabella, M.J. Gonzalez, O. Blume, A. Fehske, IEEE Wirel. Commun. **18**(5), 40 (2011)
81. T.D. Novlan, R.K. Ganti, A. Ghosh, J.G. Andrews, IEEE Trans. Wirel. Commun. **10**(12), 4294 (2011)
82. S.H. Chang, H.G. Park, S.H. Kim, J.P. Choi, IEEE Trans. Commun. **66**(11), 5828 (2018)
83. V. Chamola, B. Sikdar, IEEE Trans. Commun. **64**(12), 5278 (2016)
84. V. Chamola, B. Sikdar, IEEE Trans. Sustain. Energy **6**(4), 1650 (2015)
85. R.H.M. Zargar, M.H. Yaghmaee Moghaddam, IEEE Trans. Sustain. Energy **11**(2), 736 (2020)
86. D. Renga, M. Meo, IEEE Trans. Green Commun. Netw. **3**(2), 366 (2019)
87. M. Zheng, P. Pawelczak, S. Stanczak, H. Yu, IEEE Commun. Lett. **17**(6), 1092 (2013)
88. M. Dayarathna, Y. Wen, R. Fan, IEEE Commun. Surv. Tutor. **18**(1), 732 (2015)
89. S. Savazzi, V. Rampa, S. Kianoush, M. Bennis, IEEE Trans. Green Commun. Netw. **7**, 248–264 (2023)
90. Z. Garroussi, A.W. Badirou, M. D'Amours, A. Girard, B. Sansò, IEEE Trans. Green Commun. Netw. **7**(2), 744 (2023)
91. Q. Fan, N. Ansari, IEEE Trans. Green Commun. Netw. **2**(4), 890 (2018)
92. P. Gorla, A. Deshmukh, S. Joshi, V. Chamola, M. Guizani, IEEE Trans. Netw. Serv. Manag. **19**(3), 2714 (2022)
93. T. Han, N. Ansari, IEEE/ACM Trans. Netw. **24**(2), 1038 (2016)
94. L.A. Fletscher, L.A. Suárez, D. Grace, C.V. Peroni, J.M. Maestre, IEEE Trans. Netw. Serv. Manag. **16**(1), 279 (2019)
95. N. Sharma, K. Kumar, IEEE Trans. Netw. Serv. Manag. **20**(2), 1884 (2023)
96. C. Xu, J. Wang, Z. Zhu, D. Niyato, IEEE Trans. Netw. Serv. Manag. **16**(2), 563 (2019)
97. H. Jiang, S. Yi, L. Wu, H. Leung, Y. Wang, X. Zhou, Y. Chen, L. Yang, IEEE Trans. Netw. Serv. Manag. **15**(1), 156 (2018)
98. A. Kaur, K. Kumar, IEEE Trans. Netw. Serv. Manag. **19**(2), 1554 (2022)

99. G. Sun et al., Future Gener. Comput. Syst. **104**, 60 (2020)
100. G. Sun, G.O. Boateng, D. Ayepah-Mensah, G. Liu, IEEE Access **7**, 63588 (2019)
101. M.Y. Abdelsadek, A.U. Chaudhry, T. Darwish, E. Erdogan, G. Karabulut-Kurt, P.G. Madoery, O.B. Yahia, H. Yanikomeroglu, IEEE Trans. Commun. **71**(2), 949 (2022)
102. Bhola, Y.J. Chen, A. Balakrishnan, S. De, L.C. Wang, *Proc. IEEE VTC* (2024), pp. 1–5
103. A. Balakrishnan, S. De, L.C. Wang, *Proc. IEEE ICC* (2024), pp. 2847–2852
104. A. Balakrishnan, P. Popineau, P. Martins, arXiv preprint arXiv:2506.08739 (2025)
105. T. Song, D. Lopez, M. Meo, N. Piovesan, D. Renga, arXiv preprint arXiv:2307.00969 (2023)
106. Y. Xing, F. Hsieh, A. Ghosh, T.S. Rappaport, *Proc. IEEE VTC-Spring* (2021), pp. 1–6
107. G. Karabulut Kurt, M.G. Khoshkholgh, S. Alfattani, A. Ibrahim, T.S.J. Darwish, M.S. Alam, H. Yanikomeroglu, A. Yongacoglu, IEEE Commun. Surv. Tutor. **23**(2), 729 (2021)
108. Orange and vodacom create a joint venture to expand network coverage in rural areas in the drc. https://newsroom.orange.com/orange-and-vodacom-create-a-joint-venture-to-expand-network-coverage-in-rural-areas-in-the-drc/
109. H. Holtkamp, G. Auer, V. Giannini, H. Haas, IEEE Commun. Lett. **17**(11), 2033 (2013)
110. A. Papoulis, S.U. Pillai, *Probability, Random Variables, and Stochastic Processes* (Tata McGraw-Hill Education, New Delhi, 2002)
111. N. Kwak, C.H. Choi, IEEE Trans. Pattern Anal. Mach. Intell. **24**(12), 1667 (2002)
112. System Advisor Model: National Renewable Energy Laboratory. https://www.sam.nrel.gov
113. What will happen to solar panels after their useful lives are over? https://www.greenbiz.com
114. Y. Zhang, L. Budzisz, M. Meo, A. Conte, I. Haratcherev, G. Koutitas, L. Tassiulas, M.A. Marsan, S. Lambert, *Proc. IEEE TIWDC* (2013), pp. 1–6
115. A. Balakrishnan, S. De, L.C. Wang, IEEE Trans. Netw. Serv. Manag. **21**(3), 2888 (2024)
116. A. Balakrishnan, S. De, L.C. Wang, IEEE Trans. Sustain. Comput. **10**(3), 503 (2025)
117. W. Chen, X. Lin, J. Lee, A. Toskala, S. Sun, C.F. Chiasserini, L. Liu, IEEE J. Sel. Areas Commun. **41**(6), 1592 (2023)
118. A. Balakrishnan, S. De, L.C. Wang, *Proc. IEEE COMSNETS* (2024), pp. 391–393
119. R. Shinkuma, N. Kishi, K. Ota, M. Dong, T. Sato, E. Oki, IEEE Netw. **35**(6), 98 (2021)
120. J.M.B. da Silva, G. Wikström, R.K. Mungara, C. Fischione, IEEE Wirel. Commun. **28**(1), 44 (2021)
121. T. Yang, F. Heliot, C.H. Foh, IEEE Commun. Mag. **53**(11), 175 (2015)
122. A.K. Kiani, N. Ansari, IEEE Commun. Mag. **55**(5), 170 (2017)
123. M.J. Farooq, H. Ghazzai, A. Kadri, H. ElSawy, M.S. Alouini, IEEE Trans. Commun. **65**(2), 918 (2016)
124. C.E. Kement, F. Kara, W. Jaafar, H. Yanikomeroglu, G. Senarath, N.D. Dao, P. Zhu, IEEE Commun. Mag. **61**(7), 150 (2023)
125. N. Kato, Z.M. Fadlullah, F. Tang, B. Mao, S. Tani, A. Okamura, J. Liu, IEEE Wirel. Commun. **26**(4), 140 (2019)
126. H. Li, K. Ota, M. Dong, IEEE Netw. **37**(2), 154 (2023)
127. Antennae Location Methodology for a Telecomm Operator in India, IIM-B, Working paper no. 454. https://www.iimb.ac.in
128. N. Varshney, S. De, IEEE Trans. Commun. **69**(1), 544 (2021)
129. 3GPP: A Global Initiative. https://www.3gpp.org/specifications
130. Technical Specification Group, Radio Access Network Meeting 99427. https://www.3gpp.org/ftp
131. A. Goldsmith, *Wireless Communications* (Cambridge University Press, Cambridge, 2005)
132. Economic Times-Solar Panel Cost: price range of different types of solar panels and how much govt. subsidy you can avail for installing one. May 2014. https://economictimes.indiatimes.com
133. Batteries for solar systems-Renewable Energy Batteries. https://www.thesolarbiz.com/batteries.html
134. True Cost of Providing Energy to Telecomm Towers in India. https://www.gsma.com
135. Economic Times-Falling solar power cost lead to pricing disputes. Nov. 2009. https://economictimes.indiatimes.com

136. D. Renga, M. Meo, IEEE Commun. Mag. **60**(9), 75 (2022)
137. N. Cheng, W. Quan, W. Shi, H. Wu, Q. Ye, H. Zhou, W. Zhuang, X. Shen, B. Bai, IEEE Wirel. Commun. **27**(1), 178 (2020)
138. M.M. Azari, S. Solanki, S. Chatzinotas, O. Kodheli, H. Sallouha, A. Colpaert, J.F. Mendoza Montoya, S. Pollin, A. Haqiqatnejad, A. Mostaani, E. Lagunas, B. Ottersten, IEEE Commun. Surv. Tutor. **24**(4), 2633 (2022)
139. Q. Ren, O. Abbasi, G.K. Kurt, H. Yanikomeroglu, J. Chen, IEEE Trans. Wirel. Commun. **22**(12), 8814 (2023)
140. S. Seng, G. Yang, C. Luo, X. Li, H. Ji, *Proc. IEEE ICC* (2022), pp. 1–6
141. M. Meo, D. Renga, F. Scarpa, *Proc. IEEE GLOBECOM* (2022), pp. 1727–1733
142. J.M. Freeman, N.A. DiOrio, N.J. Blair, T.W. Neises, M.J. Wagner, P. Gilman, S. Janzou, System advisor model (sam) general description (version 2017.9. 5). Tech. rep., National Renewable Energy Lab.(NREL), Golden (2018)
143. 3GPP: Study on New Radio (NR) to support non-terrestrial networks. 3rd Generation Partnership Project (3GPP), TR 38.811, Sep. 2020. https://portal.3gpp.org/
144. G.B. Koç, B. Çiloğlu, M. Öztürk, H. Yanikomeroglu, arXiv preprint arXiv:2304.08620 (2023)
145. W.Y. Chiu, J.T. Hsieh, C.M. Chen, IEEE Trans. Ind. Inform. **16**(3), 1811 (2020)
146. Frequently Asked Questions (FAQS): How much electricity does an American home use? https://www.eia.gov/
147. D. Yang, H. Latchman, D. Tingling, A.A. Amarsingh, IEEE Potentials **34**(4), 11 (2015)
148. Y. Guo, M. Pan, Y. Fang, IEEE Trans. Parallel Distrib. Syst. **23**(9), 1593 (2012)
149. M. Nistor, C.H. Antunes, IEEE Trans. Smart Grid **9**(1), 240 (2018)
150. A. Safdarian, M. Fotuhi-Firuzabad, M. Lehtonen, IEEE Trans. Smart Grid **7**(4), 1836 (2016)
151. O. Alrumayh, K. Bhattacharya, IEEE Trans. Smart Grid **10**(6), 6284 (2019)
152. R. Khezri, A. Mahmoudi, M.H. Haque, IEEE Trans. Ind. Appl. **56**(5), 5319 (2020)
153. S. Lee, D. Whaley, W. Saman, Energy Procedia **62**, 91 (2014). 6th International Conference on Sustainability in Energy and Buildings, SEB-14
154. L. Xiao, H. Zhang, Y. Xiao, X. Wan, S. Liu, L.C. Wang, H.V. Poor, IEEE Trans. Wirel. Commun. **19**(1), 423 (2020)
155. H. Zhang, M. Min, L. Xiao, S. Liu, P. Cheng, M. Peng, *2018 IEEE Global Communications Conference (GLOBECOM)* (2018), pp. 1–6
156. R. Dubey, A. Balakrishnan, S. De, *2024 IEEE 100th Vehicular Technology Conference (VTC2024-Fall)* (2024), pp. 1–5
157. B. Mao, F. Tang, Y. Kawamoto, N. Kato, IEEE Commun. Surv. Tutor. **24**(1), 210 (2021)
158. Y. Li, S. Wang, *2018 IEEE International Conference on Edge Computing (EDGE)* (2018), pp. 66–73
159. S. Wang, Z. Liu, Z. Zheng, Q. Sun, F. Yang, *2013 International Conference on Parallel and Distributed Systems* (2013), pp. 102–109
160. A. Ibrahim, M. Noshy, H.A. Ali, M. Badawy, IEEE Access **8**, 81747 (2020)
161. M. Dorigo, M. Birattari, T. Stutzle, IEEE Comput. Intell. Mag. **1**(4), 28 (2006)
162. C. Li, W. Liu, L. Wang, M. Li, K. Okamura, J. Netw. Comput. Appl. **58**, 241 (2015)
163. C.H. Liao, J.M. Wu, J. Du, L. Zhao, *2017 9th International Conference on Wireless Communications and Signal Processing (WCSP)* (2017), pp. 1–6
164. D. Gong, J. Sun, Z. Miao, IEEE Trans. Evol. Comput. **22**(1), 47 (2018)
165. R. Zhang, Y. Chen, B. Dong, F. Tian, Q. Zheng, IEEE Access **7**, 121360 (2019)
166. L. Dai, H. Zhang, IEEE Access **8**, 83375 (2020). https://doi.org/10.1109/ACCESS.2020.2990631
167. S. Chatterjee, A.S. Hadi, *Regression Analysis by Example* (John Wiley & Sons, New York, 2015)
168. J.M. Keller, M.R. Gray, J.A. Givens, IEEE Trans. Syst. Man Cybern. **SMC-15**(4), 580 (1985)
169. Z.M. Fadlullah, F. Tang, B. Mao, N. Kato, O. Akashi, T. Inoue, K. Mizutani, IEEE Commun. Surv. Tutor. **19**(4), 2432 (2017)
170. R. Dong, C. She, W. Hardjawana, Y. Li, B. Vucetic, IEEE Trans. Wirel. Commun. **20**(4), 2309 (2021)

171. S. Wang, M. Chen, C. Yin, W. Saad, C.S. Hong, S. Cui, H.V. Poor, IEEE Internet Things J. **8**(24), 17460 (2021)
172. J. Tang, X. Li, H. Li, P. Li, X. Wang, V.C.M. Leung, IEEE Trans. Mob. Comput. **24**(7), 5681 (2025)
173. S. Zhou, L. Feng, M. Mei, M. Yao, IEEE Internet Things J. **11**(18), 30400 (2024)
174. M. Liaq, W. Ejaz, *ICC 2024 - IEEE International Conference on Communications* (2024), pp. 1292–1297
175. Z. Zhao, M. Feng, C. Ke, Z. Chen, T. Jiang, IEEE Internet Things J. **11**(15), 26444 (2024)
176. Z. Li, Z. Lin, J. Shao, Y. Mao, J. Zhang, IEEE Trans. Mob. Comput. **23**(11), 10509 (2024)
177. C.C. González, E.F. Pupo, E. Iradier, P. Angueira, M. Murroni, J. Montalban, IEEE Trans. Veh. Technol. **73**(8), 11862 (2024)
178. J. Xue, Z. Qu, J. Xu, Y. Liu, Z. Lu, IEEE Trans. Mob. Comput. **23**(6), 7470 (2024). https://doi.org/10.1109/TMC.2023.3335258
179. C.H. Liu, Z. Chen, J. Tang, J. Xu, C. Piao, IEEE J. Sel. Areas Commun. **36**(9), 2059 (2018)
180. S. Khairy, P. Balaprakash, L.X. Cai, Y. Cheng, IEEE J. Sel. Areas Commun. **39**(4), 1101 (2021)
181. D.A. Temesgene, M. Miozzo, D. Gündüz, P. Dini, IEEE Trans. Sustain. Comput. **6**(4), 626 (2021)
182. C. He, Y. Hu, Y. Chen, B. Zeng, IEEE J. Sel. Areas Commun. **37**(10), 2200 (2019)
183. M. Simsek, M. Bennis, I. Guvenc, IEEE Trans. Veh. Technol. **64**(10), 4589 (2015)
184. L. Zhang, Y.C. Liang, IEEE Trans. Wirel. Commun. **20**(4), 2551 (2021)
185. S. Gong, Y. Xie, J. Xu, D. Niyato, Y.C. Liang, IEEE Netw. **34**(5), 106 (2020)
186. S.J. Pan, Q. Yang, IEEE Trans. Knowl. Data Eng. **22**(10), 1345 (2010)
187. S. Sharma, S.J. Darak, A. Srivastava, *2019 URSI Asia-Pacific Radio Science Conference (AP-RASC)* (2019), pp. 1–4
188. S. Sharma, S.J. Darak, A. Srivastava, *2017 9th International Conference on Communication Systems and Networks (COMSNETS)* (2017), pp. 397–398
189. C. Pradhan, A. Li, C. She, Y. Li, B. Vucetic, IEEE Trans. Commun. **68**(7), 4565 (2020)
190. M. Assran, Q. Duval, I. Misra, P. Bojanowski, P. Vincent, M. Rabbat, Y. LeCun, N. Ballas, *Proceedings of the IEEE/CVF Conference on Computer Vision and Pattern Recognition (CVPR)* (2023), pp. 15619–15629
191. A. Bardes, J. Ponce, Y. LeCun. Mc-jepa: a joint-embedding predictive architecture for self-supervised learning of motion and content features (2023). https://arxiv.org/abs/2307.12698
192. C.C.J. Kuo, A.M. Madni, J. Vis. Commun. Image Represent. **90**, 103685 (2023). https://doi.org/10.1016/j.jvcir.2022.103685. https://www.sciencedirect.com/science/article/pii/S104732032200205X

The manufacturer's authorised representative in the EU is Springer Nature Customer Service Centre GmbH, Europaplatz 3, 69115 Heidelberg, Germany. If you have any concerns regarding our products, please contact ProductSafety@springernature.com

Printed and bound by CPI Group (UK) Ltd, Croydon, CR0 4YY
07/10/2026
02229361-0005